The Anunnaki Series

Anunnaki, UFOs, Extraterrestrials And Afterlife Greatest Information As Revealed By Maximillien de Lafayette.
Part 3.
4th Edition

Selections from his 50 years of studying with Anunnaki Ulema, his secret findings & his writings

A set of 3 Books

*** *** ***

Copyright ©2010 by Maximillien de Lafayette. All rights reserved. No part of this book may be used or reproduced by any means, graphic, electronic, or mechanical, including photocopying, recording, taping or by any information storage retrieval system without the written permission of the author except in the case of brief quotations embodied in critical articles and reviews.

Acknowledgment and Gratitude

I am deeply grateful to the Honorable Anunnaki-Ulema who have generously contributed to this book.
My teachers:
Master Li
Ulema Sharif Al Mutawalli
Ulema Mordachai Ben Zvi
Ulema Sadik Bin Jaafar Al Kamali
Ulema F. Oppenheimer
Ulema A. Berkof
Ulema Ramash Govinda
Ulema Tabeth Al-Baydani
Ulema Shaul Sorenztein
Sinhar Ambar Anati
Ulema F. Tayara
Ulema Mirach Faridi Beraz
For without their guidance and contributions, this book would have remained stacks of papers in my drawer.

A special thankyou note goes to my wonderful friends, **Laura Lebron,** and **Melinda Pomerleau**, who gave me so much energy that enables me to complete this series. They are the best of the best people I have ever met in this life...

Anunnaki, UFOs, Extraterrestrials And Afterlife Greatest Information As Revealed By Maximillien de Lafayette.

Part 3. 4th Edition
Selections from his 50 years of studying with Anunnaki Ulema, his secret findings & his writings

Condensed from the previously published two volume series
The Best of Maximillien de Lafayette: UFOs, Anunnaki, Extraterrestrials
Knowledge Not Found Elsewhere

A set of 3 Books

Maximillien de Lafayette

Times Square Press. Elite Associates
New York   London   Paris

*** *** ***

2010

Table of Contents of Part 3

Chapter 42
Female extraterrestrial goddesses created us, not Jehovah, God or Allah...25

- Female extraterrestrial goddesses created us, not Jehovah, God or Allah...25
- The Bible, religious texts and the Scriptures were the product of religious visions rather than reality and truth...25
- The universe and the human race were not created by Jehovah, God and Allah...26
- The extraterrestrial goddesses who created the human race...27
- The Mother Goddess was considered the life giver and the creator of all life-forms, including mankind...27
- Angel Gabriel...28
- Gabriel was a woman, not a man, because she was described as "the female who created life...28
- Gabriel as a female Anunnaki was the first to experiment with copies of a human, later to be called Adam...29
- Gabriel created 7 different types of Homo Sapiens by using the DNA of primitive beings and the DNA of an Anunnaki...29
- Adam, the man, was first created from the rib of Gabriel, the female Anunnaki known to the Sumerians as the "lady of the rib" ...29
- Aruru...30
- Aruru, the Sumerian goddess of earth who created the first specie of man from clay found in Iraq...30

- Athirat created the gods and the human race in Phoenicia...31
- Dates of the Creation according to the Bible, Koran, Sumerians, Chinese, Egyptians...31
- Avalanches of "Extraterrestrial Mothers" of the universe who fashioned our world and genetically created us...33
- Aa "Ai"...34
- Anunnaki's women or goddesses were the creators of the human race...34

Chapter 43
Amram...35

- What is Amram? ...35
- Linked to Enki, Ea and Abraham...35
- Ram acquired a multitude of meanings...35

Chapter 44
An'th-Khalka...39

- What is An'th-Khalka? ...39
- Excerpts from Ulema's Kira'at (Reading) ...39
- Mankind was not created by one single Sumerian god...39
- A multitude of gods and goddesses created different types and categories of human beings...40
- Ninlil...40
- Marduk...41
- Inanna......41
- Namma...41
- Ea...42

Chapter 45
The Anunnaki as they were known to ancient tribes and civilizations...45

- The Anunnaki as they were known to ancient civilizations...45
- The real meaning of the word Anunnaki...45
- Possibly a wrong translation of the word Anunnaki on the part of authors...45
- How the word Anunnaki was used...46
- The different meanings of the word Anunnaki...46
- The Anunnaki were known to many neighboring countries in the Near East, Middle East, by and under different names...46
- Various attributes or definitions were given to the Anunnaki...47
- The Anunnaki in the Enuma Elish...48
- Epistemology and historical terminology of the word Anunnaki...48
- Linguistic examples...49

Chapter 46
Ataba-Darr-Ja and Atabukha Darja...51

- What is Ataba-Darr-Ja? ...51
- To the Anunnaki, the universe was NOT created by God...51
- What is Atabukha Darja? ...51

Chapter 47
The Anunnaki view of the Afterlife...53

- Humans should not be afraid to die, nor fear what is going to happen to them after they die...55
- Soul is an invention of mankind. It does not exist anywhere inside us. Instead, there is a non-physical substance called "Fik'r" ...56
- What is Fik'r? ...56
- The "Fik'r" is linked to the Anunnaki our creators through a "Conduit" found in the cells of the brain...56
- We are a copy of an original copy of ourselves stored in the "Rouh-Plasma"; a compartment under the control of the Anunnaki...57
- You, your so-called "soul", your mind and your new shape in the Anunnaki's Bubble...57
- Is the afterlife a physical world? ...58
- Everything you have experienced on earth exists in other dimensions, and there are lots of them...58

Chapter 48
All That Is...61

- To the Anunnaki, God is made of inexhaustible mental energy...61
- The possibility exists that there are many Gods...61
- The Anunnaki are indeed the creators of human beings...62

Chapter 49
Sexual relationships and reproduction of the Anunnaki and extraterrestrials...65

- Anunnaki reproduction is done by technology, involving the light passing through the woman's

- body until it reaches her ovaries and fertilizes her eggs...65
- Anunnaki's sex involves an emanation of light from both participants...65
- Like many extraterrestrials, the Anunnaki do not have genital organs...66
- Major points about the subject ...66
- Aliens reproduce in laboratories...66
- Alien babies are nourished by a "light conduit." ...66

Chapter 50
Immaculate Conception; a typical Anunnaki pattern...71

- Cherchez the Bible! ...71
- Impregnated women by "something" or "somebody" other than their husbands...71
- Got pregnant before they were married...71
- Almost every famous and notable female in the Bible had some sort of a mind-boggling pregnancy situation...72
- Here are some of the famous women of the Bible who got pregnant by something or somebody else than their husband...73
- Pregnant before marriage...73
- Very old and barren women...73
- Desperately old women...73
- Barren until middle age...73
- Famously barren...73
- Was David an Anunnaki offspring? ...74
- In the Bible, so many of the important female characters are barren and unable to conceive –

until a very special person comes, sometimes as an angel, to help...75

Chapter 51
Al Jahiliya...79

- What is Al Jahiliya? ...79
- But how does this fit in the Anunnaki scenario? ...79
- The Raheb Bouhayra, Prophet Mohammad, Islam and Anunnaki connections...80
- The practice of the occult and the study of extraterrestrial literature were abolished by Islam...81
- The society of Ulema of Rama-Dosh...81

Chapter 52
The Anunnaki's Ra-Nif...83

- What is Ra-Nif?...83
- The many different forms of spirits...83
- The union between the Anakim or Anaki and the daughters of men (Women of the earth) created a new race of beasts and demons...83

Chapter 53
Kefitzat ha-derekh...87

- What is kefitzat ha-derekh? ...87
- The Talmud, however, mentions three Biblical individuals who experienced it...87

- Kefitzat ha-derekh can happen spontaneously, as a miracle performed for the benefit of a just and good person who is in trouble...87
- Such a miracle is assumed to be performed either by God himself, by one of his angels, or by Elijah the Prophet...88
- The secret knowledge of the holy names...88
- Baal shem and Baal Shem Tov...89
- Kefitzat ha-Derekh is only one of the many wonders the Baalei Shem...89
- The concept appeared in many areas of the world...90
- The idea resembles teleportation in other world myths and legends...91

Chapter 54
On the road to ultimate knowledge...95

- Excerpts from the book "On the Road to Ultimate Knowledge: Extraterrestrial Tao of the Anunnaki and Ulema...95
- Meeting the master, and mastering fears...95
- Scene: The little Germain is walking with his mother in the streets of Paris...95
- Year: 1945, right at the end of World War Two. France has been liberated by the Americans...95
- A group of six or seven, wearing black clothes and berets and carrying large, very visible guns, were dragging a young girl...95
- All during the war my room was the only place where I felt entirely safe...97
- Meeting my future Master...99
- In the living room stood a person that seemed to me like a materialization of a cartoon...101
- The first miracle...103

- Leaving Paris…104
- Arriving to Benares…105
- My first encounter with the paranormal…105
- And the snake obeys the Master…105
- In the house of the Master…106
- Temples, flowers, and lit oil lamps, to float on the river…106
- The magical room of colored papers…107
- And then came the miracle of the tree…108
- Miracles and tricks of illusion in the streets of India…109
- Return to Paris…112
- On the road again…on my way to Kowloon…113
- The mirage island…114
- The small temple on the top of a hill…115
- Talking to the Buddha…115
- The villagers, poor as they were and experiencing constant struggle for survival, were unusually loving, and happy…117
- Feeding the fish with food from heaven…118
- The secret is how to use your hands properly…119
- The Master: "You will find that animals understand more than people do, if you love them." …120
- Ballet with nature…121

Chapter 55
The unthinkable horrors of the Grays…125

- Victoria (Sinhar Ambar Anati) in her own words…125
- How I was taught the truth about alien abductions…125
- The unthinkable horrors…125
- What do the Grays want of us on earth? …127

- The Anunnaki monitor...127
- The room was full of operation tables, and human beings, each attached to the table and unable to move...128
- The needles reached every part of the human bodies, faces, eyes, ears, genitals, stomach...129
- The Grays can enter the human mind quite easily...129
- Most of the Grays have no substance, and therefore they are invisible...131
- "This is a baby's head! A floating baby's head! What in the world it is attached to?" ...132
- Many of the abductees who are thus psychologically influenced fall into a pathological attachment to the hybrid child...134
- Some look reptilian, some insect-like. Some eat human beings and other sentient species...134
- At the Grays' hospital ward: Inside floated a large number of severed arms and legs, all human...136
- Each tank filled with body parts, some even with full bodies...136
- The smell of formaldehyde was so intense that I almost fainted...137

Chapter 56
At the Anunnaki's Academy in Nibiru...143

- At the Anunnaki's academy...143
- In Victoria's own words...143
- Anunnaki's orientation program...143
- The lessons are entered directly into the brain's cells of the students...144
- The Anunnaki beautiful classroom...145
- The Nif-Malka-Roo'h-Dosh ritual...149
- The creation of the mental Conduit...151

- The "Double" and "Other Copy" of the mind and body of the student...151
- The Anunnaki have collective intelligence and individual intelligence...152

Chapter 57
On extraterrestrials currently living among humans and USO...155

- Categories of extraterrestrials living among us...155
- Extraterrestrial Grays from various stars and planets...155
- Hybrids who are half humans-half aliens...155
- Extraterrestrials from various spatial races who live underwater...155
- The USO...156
- Many reports were submitted by sea captains and authenticated by military crews...156
- The recorded USOs' reports are in the thousands...157
- The "Malta Water-UFO/USO...157
- The "American UFO/USO-spy plane and metal memory substance...157
- The Argentinian 500 miles per hour USO...158
- Lake District, United Kingdom's USO sightings...159
- Laguna Cartegena USO activities...159
- But why underwater bases? ...159
- Ideal locations for extraterrestrials...159
- Are they harmful to us? ...160
- They are not, because they are from our future...160
- Extraterrestrials cannot alter the past...160
- Essential and unique information about USO...160

Chapter 58
Key words and important terminologies from the Anunnaki language and ancient Middle Eastern epics/tablets...169

- What or who is Aa?...169
- Who is Abd?...170
- The genetic composition of Abel...170
- Eve's relation to the Anunnaki...171
- The concept of Eve conceiving Cain and Abel with the help of God...171
- What is Abra-ah? ...172
- What is Afarit? ...172
- What is Amram? ...173
- Abraham, God and the Anunnaki face to face...173
- What is An'h-Ista-Khan-na-reh? ...174
- What is Anda-anparu? ...175
- What is An-Maha-Rit? ...175
- What is An-Nafar JinMarkah? ...176
- What is Anunnaki ME.nou-Ra "MEnour"? ...177
- What is Anunnaki.Nou-Rah-Shams."Menora-Shems"? ...177
- What is An-zalubirach? ...178
- What is Anzuma? ...178
- What is Arakh-nara "Arcturus"? ...178
- Edgar Cayce did mention it...179
- Arcturus is the first docking station that allows us to travel beyond our consciousness...179
- What is Arwad "Aradus"? ...179
- Arwad hold many great secrets, to name a few...180
- What is Ay'inbet? ...181
- Baalshalimroot...181
- Those who have practiced As Sihr (Magic) used their eyes as a psychic conduit...182

- The concept of hell was unknown to the pagan Badou Rouhal...183
- Esoteric benefits...183
- What is Azaridu? ...183
- A selection of important words from the Anunnaki's language...184
- Aamala...184
- Abamarash...184
- Abekira'h-Kitbu...184
- Aberuchimiti...184
- Abgaru...185
- Abra-ah...185
- Akama-ra...185
- Anšeduba...185
- Anubdada...186
- An-zalubirach...187
- Anzuma...187
- Baliba nahr usu na Ram...188
- Balu-ram-haba...188
- Barak yom-ur...190
- Barak-malku...190
- Chavad-nitrin...191
- Chimiti...191
- Dab'Laa...191
- Dadmim "Admi", "Adamai", "Adami" ...192
- Da-Irat...192
- Dayyakura...197
- Dhikuru...197
- Eido-Rah...197
- Emim...198
- Ezakarerdi, "E-zakar-erdi "Azakar.Ki" ...199
- Ezakarfalki "E-zakar-falki"...200
- Ezbahaiim-erdi...200
- Ezrai-il...200
- F "ف" ...201

- Fari-narif "Fari-Hanif" ...201
- Ghen-ardi-vardeh "Gen-adi-warkah" ...201
- Gens "Jenesh" ...205
- Gensi-uzuru...205
- Ghoolim...206
- Giabiru...207
- Gibbori...207
- Gibishi...207
- Gibsut-sar...207
- Gilgoolim...208
- Ginidu...208
- Girzutil...208
- Goirim-daru...208
- Golibu "Golibri" ...208
- Golim...208
- Gomari "Gumaridu...209
- Gomatirach-minzari "Gomu- minzaar"...213
- Prerequisites...215
- Precautions...216
- Equipment and supplies...217
- Building the Minzar...218
- Contacting the alternate realities...219
- Subsequent visits to the alternate realities...222
- Benefits and advantages...222
- Returning to your regular reality on Earth...223
- Gubada-Ari...224
- Synopsis of the theory...225
- Materials...225
- The technique...226
- Gudinh...232
- Hag-Addar...232
- Hamnika-mekhakeh...232
- Hamnika-mekhakeh- ilmu...232
- Synopsis of the concept...232
- The Ulema-Anunnaki days are...232

- The calendars' grids...233
- The use of a language...233
- The preparation and use of the grids...234
- Grid 1: Calendar of the week...234
- Grid 2: Calendar of your name...235
- Calendar of your lucky hour...236
- Grid 4...236
- Grid 5...237
- Hamsha-uduri "Dudurisar" ...238
- The concept...238
- The technique: It works like this...239
- How real is the holographic/parallel dimension you are visiting? ...245
- Some of the benefits...247
- Closing the technique...247
- Harabah...248
- Haridu "Haridu-ilmu" ...248
- Harima...251
- Harimu...251
- Harranur-urdi...251
- Functions of the Conduit, Miraya and retrieving data...254
- Hatani...262
- The concept...262
- The Hatani protection shield...264
- Hatori "Hatori-shabah" ...266
- Hattari...267
- Hawaa...289
- Hawwah...289
- Hawwah/Eve: Transliteration of a text from the Book of Ramadosh...291
- Translation of the text from the Book of Ramadosh...292
- Haya-Saadiraat...293
- Hazi-minzar "Mnaizar" ...293

Chapter 42
Female extraterrestrial goddesses created us, not Jehovah, God or Allah

- Female extraterrestrial goddesses created us, not Jehovah, God or Allah...25
- The Bible, religious texts and the Scriptures were the product of religious visions rather than reality and truth...25
- The universe and the human race were not created by Jehovah, God and Allah...26
- The extraterrestrial goddesses who created the human race...27
- The Mother Goddess was considered the life giver and the creator of all life-forms, including mankind...27
- Angel Gabriel...28
- Gabriel was a woman, not a man, because she was described as "the female who created life...28
- Gabriel as a female Anunnaki was the first to experiment with copies of a human, later to be called Adam...29
- Gabriel created 7 different types of Homo Sapiens by using the DNA of primitive beings and the DNA of an Anunnaki...29
- Adam, the man, was first created from the rib of Gabriel, the female Anunnaki known to the Sumerians as the "lady of the rib" ...29
- Aruru...30
- Aruru, the Sumerian goddess of earth who created the first specie of man from clay found in Iraq...30
- Athirat created the gods and the human race in Phoenicia...31

- Dates of the Creation according to the Bible, Koran, Sumerians, Chinese, Egyptians...31
- Avalanches of "Extraterrestrial Mothers" of the universe who fashioned our world and genetically created us...33
- Aa "Ai"...34
- Anunnaki's women or goddesses were the creators of the human race...34

*** *** ***

Chapter 42

Female extraterrestrial goddesses created us, not Jehovah, God or Allah

> We were taught that God (The Judeo-Christian-Muslim God) created the world and the human race. Yet, the Bible (Old Testament), the New Testament, and the Quran (Koran) failed to produce one single evidence. Our religions (Judaism, Christianity and Islam) are based upon faith. But what is faith anyway? Faith is what we get when we don't have proof. Faith is a fantastic invention by mankind much needed to bring comfort, hope and consolation. But why faith has to substitute for truth? Faith is a very weak argument.

The Bible, religious texts and the Scriptures were the product of religious visions rather than reality and truth

As science explores new frontiers of knowledge, and the human brain begins to understand how the universe was created, how stars and planets, including planet earth came to inhabit the landscape of the cosmos, and how light, gas, oxygen, carbon and cosmic dust created the human body, we, as humans occupying a minimal and insignificant fraction of this cosmic landscape, we begin to realize that:

- Ancient religious texts were the product of religious visions rather than reality and truth.
- The sacred books of our religions were crafted by nomads, by tribes that lived in the desert, in uncultured and uneducated communities, in savage habitats, by religious zealots wearing animal furs and camels' skins, and by brilliant story-tellers who had access to fabulous stories, legends and myths written centuries and centuries and centuries by much older tribes, religious visionaries and communities.

- Many of the Biblical stories were taken word for word from Babylonian, Assyrian, Egyptian, Sumerian and Phoenician epics, and rewritten according to the needs and raison d'etre of new religious.
- Almost all the most meaningful and important episodes, accounts, stories and events of the Bible were already written and recorded in Sumerian, Babylonian, Egyptian and Phoenician epics, myths, scrolls, and on clay tablets. Some existed thousands of years before Jehovah, God, Jesus, Islam, Mohammad and the Quran became known to mankind.
- The founding fathers of our religions created a patriarchal society, a patriarchal religion, and patriarchal laws governing all facets and aspects of our daily life, social order, justice, and even the gender of the creator of the universe.

The universe and the human race were not created by Jehovah, God and Allah

If we go to the source of the sacred religious texts and scriptures, we will find out that:

A-The universe and the human race were NOT created by Jehovah, God and Allah.
B-God is not masculine.
C-God as we know him today never spoke to Abraham, Moses, Jesus and Mohammad.
D-The earliest records of humanity history, and the civilizations of mankind provided ample explanations and irrefutable documentation on the origin of the human race, the primitive humans, their habitats, their way of life, their diet, their migration patterns, and above all the events that shaped earth and man's understanding of the "divine".
E-The human race was created by advanced extraterrestrial races that descended on earth – first in Phoenicia, and second in Sumer– some 450,000 years ago.

*** *** ***

The extraterrestrial goddesses who created the human race

The Mother Goddess was considered the life giver and the creator of all life-forms, including mankind.

In the beginning, at the very dawn of the religions in the Near East, and the Middle East (Lebanon known as Phoenicia, Syria, Palestine/Israel, Isle of Arwad, Cyprus, Jordan, Iraq, Sinai), the Mother Goddess was considered the life giver and the creator of all life-forms, including mankind.

The world, the planets, animals and all humans were created by a woman; a "Mother Goddess". She was worshiped as the "Supreme Goddess of the Universe and Life" by Phoenicians, Hyksos, Canaanites, Hebrews, Akkadians, Hittites, Babylonians, Sumerians, and Egyptians et al...But later on in history, religions became dominated by male gods.

Nevertheless, female goddesses remained important, and played a paramount role in the daily life of humans.

The supreme male god became "EL", and meant "first, ultimate" in Aramaic, "Power" in early Hebrew, and the "Origin, number one and god of gods" in Phoenician. When El is used as a name, El becomes the main god, the god of all gods.

Yes, we were created by women, not men and not by God.

But who were these women, creators of the universe and the human race?

Where did they live?

How do we know they were in fact women?

Any record?

Any scripture?

Any proof? Were they known to our ancestors?

Where?

When?

And How?

Here are some answers taken directly from historical texts and original sources that created our religions:

To name a few:

1-Angel Gabriel

Gabriel was a woman, NOT a man, because she was described as the female who created life

> Angel Gabriel is a Jewish-Christian-Muslim Biblical figure. Where did he come from? And how did we learn about him? Of course, we did learn about him from the Bible, where else? Really? Not really, because Angel Gabriel existed longtime before the Bible was written. He was called Satan-il, and he was an Anunnaki personage with mighty powers and major influence on the creation of the human race. Angel Gabriel is not totally and exactly what the Judeo-Christian tradition portrays.

His original name is "Gib-ra-il"; the guardian of Janat Adan or Edin (Garden of Eden), in Sumerian, and in Anakh, Gi-ra-I of Grabriel is Nin-il, or Nin-Lil. Gabriel is also called "Gab" and "Gab-r-il". Gab means a female guardian, a governor or a protector. This explains why Angel Gabriel was represented to us as the guardian of the Garden of Eden.
In the ancient texts of the Sumerians, Akkadians and civilizations of the neighboring regions, "Gab-r" was the governor of "Janat Adan."
In various Semitic languages, "Janat" means paradise, and Eden is Idin or Adan. This is how we got Garden of Eden. Angel Gabriel, the Sumerian is more than a guardian, because he was called Nin-Ti which means verbatim: Life-Woman. In other words, Angel Gabriel was three things:

1-Governor of the Garden of Eden,
2-A woman, NOT a man, because she was described as "the female who created life",
3-A geneticist who worked on the human DNA/creation of the human race. The word "Gab-r" was phonetically pronounced as: Gab'er. The early Arabs adopted it as "Al Jaber" meaning many

things including: force, authority, might, and governing. From "Al Jaber" important words, nouns and adjectives were derived. For instance, the word "Jabbar" means: mighty, powerful, capable, huge, giant, like the giants in the Bible and Sumerian/Anunnaki epics.
"Jababira" is the plural of "Al Jabbar".
After the Arabs were converted to Islam, "Al Jaber" became "Al Jabbar"; one of the attributes and names of Allah (God). In the Anakh (Anunnaki language), the word "Jabba-r-oout" means exactly the same thing in early Aramaic and modern Arabic: Authority, power, rule, reign.

Gabriel as a female Anunnaki was the first to experiment with copies of a human, later to be called Adam

Major Points:
- Gabriel created 7 different types of Homo Sapiens by using the DNA of primitive beings and the DNA of an Anunnaki

- Adam, the man, was first created from the rib of Gabriel, the female Anunnaki known to the Sumerians as the "lady of the rib".

It is so obvious that the Anakh language deeply influenced Eastern and Western languages.
One more surprise for the readers. We find striking similarity in our Western vocabularies (Latin, Anglo-Saxon, French and Romance languages); Gab'r became gouverneur in French, governor in English, and gubernator in Latin.
The Sumerian Gabriel was also known under different names according to the Sumerian texts, such as "Nin-Hour-sagh", meaning the lady governess of the mountain; an elevated region of the Garden of Eden. Gabriel as a female Anunnaki was the first to experiment with copies of a human, later to be called Adam. But first, Gabriel created 7 different types of Homo

Sapiens by using the DNA of primitive beings and the DNA of an Anunnaki.

Gabriel's original creations were not very successful. Later on, Gabriel used a most unexpected genetic source to create the final copy of the modern man.

There are plenty of evidence and historical statements to prove this point. And all began with her name "Nin-il", some times referred to as "Nin-ti".

In Anakh, Sumerian and Babylonian languages, the word "Ti" means "rib". In later versions of the ancient texts, "Nin-ti" became the "lady of the rib", also the "lady of life", and the "lady of creation".

Consequently, Adam, the man, was first created from the rib of Gabriel, the female Anunnaki known to the Sumerians as the "lady of the rib". This of course contradicts the story of the creation of Adam and Eve as told in the Judeo-Christian tradition.

According to the Sumerians and the Anakh, a woman created man; it was not a man who created a woman (Eve). And the female Anunnaki (Gabriel) used her rib to create Adam.

The early Hebrew scribes and later, the translators of the ancient texts and epics of Sumer and Phoenicia got it wrong. Their fake stories of the creation of Adam and Eve, the Genesis, and purpose of the Creation invaded the Hebraic, Christian and Islamic Holy Scriptures.

Aruru

Aruru, the Sumerian goddess of earth who created the first specie of man from clay found in Iraq

Aruru was the tender and caring Sumerian goddess of earth who created the first specie of man from clay found in Iraq. She is also called Nintu, Ninhursaga, Mami, Beletili and Ninmah. Later on, Gb'r (Angel Gabriel) assumed the duty of creating Adam, but Aruru maintained her position of power and decisive authority in decision-making, related to the creation of new prototypes of mankind. In Anakh, her name is "Ar-Rouh", meaning "the creator of life; the giver of the spirit."

Rouh became "Rouh" in Arabic, "Rohka", and "Rouah" and "Roach" in Aramaic, and Hebrew, and all these linguistic variations share the same meaning: Spirit.

Aruru was also called Belet-ili, Ninmah, Nintu, Ninhursasga, and Mami. She was the mother goddess responsible for the creation of man with the help of the Anunnaki gods, Enlil or Enki.

Following the instructions of Ea, she mixed clay with the blood of an Igigi's god, Geshtu-e, and created seven men and seven women.

And obeying Anu's command, Aruru created Enkidu in Anu's image by using a clay, throwing it into the wilderness, and nusing him according to specific Anunnaki instructions.

Athirat created the gods and the human race in Phoenicia

There were many important goddesses in the Phoenician and Canaanite religions. According to the Phoenician-Ugaritic epic, the most prominent were Athirat, Athtart, Anath and Derketo. The Mother Goddess was superior to all the other goddesses. Athirat was the consort of God El who became the god of the Hebrews. Athirat created the gods and the human race in Phoenicia. She used her breath and earthy elements to "manufacture" human bodies, and she placed the early humans in fertile and green prairies.

Athirat also protected her creatures by assigning an angel guardian to watch over them (Déjà vu?)

Dates of the Creation according to the Bible, Quran, Sumerians, Chinese, Egyptians...

- According to the Bible (The Old Testament), the Creation happened 6000 years ago.
- The New Testament and Christianity accept this period.

- According to the Koran, and general Islamic beliefs, the Creation has occurred 5000 years ago.
- According to the Sumerians, the Creation happened 241,200 years ago;
- To the Chinese 49,000 years ago;
- To the Egyptians 13,000 years ago;
- To Herodotus 17,000 years ago.
- To the Anunnaki-Ulema, 16 billion years ago.

None of these periods are in harmony with the evidence and dates shown by the scientific studies in our age (Except the Anunnaki-Ulema's date). One has to think and think hard, what is the reason behind the differences in the dates of the Creation?
If the creator is the one, and the only one who created everything, then why these scribes and scholars don't have a single time span for the creation?

According to modern science, the time of the Creation is between 12 and 16 billion years, and with every new scientific discovery the dates keep on changing.
But one story remains the most striking: The universe and the human race were created by a goddess; a female creator, an extraterrestrial woman who reigned over heavens, the skies, earth and mankind.
The joint effort of Lahar the Cattle God, and Ashnan the Grain Goddess resulted in the creation of man. But this is not the only myth which resulted in the creation of humanity.
The Sumerian myth, the Phoenician texts, and the Babylonian Epic of Creation differ considerably. But the reason for which man was created is identical in both accounts.
In the Sumerian myths, the Anunnaki gods complain that they could not get their food. They appeal to Enki, who, was asleep. Nammu, the mother of the gods, wakes him up, and reports to him that the gods, and the Igigi are furious. Enki told Nammu and Ninmah to do something about it, and following the Anunnaki gods instruction, Nammu and Ninmah, with the help of other deities mixed clay they found over the abyss, and fashioned the first prototypes of Man.
The Phoenician and Sumerian goddesses, Ishtar and Inanna also created the "early forms" of humans by mixing clay with the DNA (Blood) of an extraterrestrial Anunnaki god.

Avalanches of "Extraterrestrial Mothers" of the universe who fashioned our world and genetically created us

In addition to these goddesses, there are avalanches of "Extraterrestrial Mothers" of the universe who fashioned our world and genetically created us. Although male gods were part of the creation process, none of them caused his final seal on the shape, form and design of the human race.
Yes, men were created by women goddesses who lived among them, and protected them, and taught them art, languages, music, science and literature ...No! mankind was not created by a vengeful god who spoke once or twice to two men somewhere in the desert (Abraham and Moses)...only two men in the entire history of humanity, and never said another word to 6 billion people who are still searching or waiting for him!!!

> **Et Dieu créa la femme. Et la femme créa le reste... (And God created the woman. And the woman created the rest...)**

*** *** ***

Aa "Ai"

Anunnaki's women or goddesses were the creators of the human race

Aa "Ai" is an Assyrian expression/term to represent the female strength of the sun. Usually written as Aa na shams. Shams means sun in Assyrian; in Arabic, it is Shams; In Hebrew, it is Shemesh. In Phoenician, it is Shama or Shem.
When Aa is used as Ai, the meaning becomes: Negative; enemy.
Nebuchadnezzar said: "Ai isi nakiri"; meaning: May I not have enemies.
Sardanapalus said: "Kasid ai-but Assur," meaning: Capturing the enemies of Assur.
In the Anunnaki's literature, women played a major role in the human affairs, as well as in the first three Anunnaki's expeditions to planet earth.
Anunnaki's women or goddesses were the creators of the human race; they were the first geneticists who produced the seven human prototypes, and adjusted the Conduit in the brains' cells of the early humans.
Anunnaki's goddesses were known to the early Phoenicians of Arwad as the sun-goddesses.
They had both positive and negative energies, depending on the intentions of each goddess. The Assyrian concept of the female strength of the sun derived from local Phoenician legends, based on oral history and tales known to the Arwadians.

*** *** ***

Chapter 43
Amram

What is Amram?
Amram is an Ana'kh word or term for the good subjects of the Anunnaki's leaders, and the good communities. It is composed of two words: Am (Good; kind) + Ram (People; community; population; tribe.)

- In Biblical studies, Amram means high people; kindred of the High; friend of Jehovah.
- In primitive Arabic, Ram meant: People; group. Henceforth, the name of the Palestinian city Ramallah could be interpreted as the people of God, since Allah means to the Arabs and Muslims, what exactly the word Jehovah means to the Jews: God.

Linked to Enki, Ea and Abraham...
When Enki or Ea called upon Avraham, he told him: I am your god, and I am now changing your name from Av-raham to Ab-Raham, because you are going to lead my people as the father of my people on earth. Av became Ab. And Ab in all the 14 different ancient languages of the Near East and the Middle East means father.
From the Ana'kh Ab, derived the words: Ab, Abu, Abi, Aba, Abba, Abuya, Abouna; all meaning the very same thing: Father. And from the Ana'kh word Ram, derived the ancient Hebrew, Aramaic and Arabic word Ram: People.

Ram acquired a multitude of meanings

Centuries later, Ram acquired a multitude of meanings.
For instance:
- **a**-In ancient Hebrew, Ram is pleasing;
- **b**-In Sanskrit mythology, Ram means supreme.

- **c**-In the pre-Islamic Arabic era, called Al-Jahiliya, (Ages of darkness), Ram meant a group of people. Synonym: Ra'bh.

From the Ana'kh Ram, we have today, the Arabic Ramy and the Spanish Ramos. In ancient times, the early Armenians called themselves the people of Ram.

The erly Armenians recognized themselves as the People of the Ram and their supreme deity was Khal-di.

*** *** ***

Chapter 44
An'th-Khalka

- What is An'th-Khalka? ...39
- Excerpts from Ulema's Kira'at (Reading) ...39
- Mankind was not created by one single Sumerian god...39
- A multitude of gods and goddesses created different types and categories of human beings...40
- Ninlil...40
- Marduk...41
- Inanna......41
- Namma...41
- Ea...42

*** *** ***

Chapter 44
An'th-Khalka

What is An'th-Khalka?

An'th-Khalka is an Ana'kh and Ulemite term composed of An'th (Race; people) + Khalka (Creation; birth). General meaning: Creation of the first humans. From An'th derived the Arabic word Ounth, which means people, humans, human race.
And from Khalka, derived the Arabic word Khalika, which means creation of the human beings.

Excerpts from Ulema's Kira'at (Reading):

Mankind was not created by one single Sumerian god

- More than one Anunnaki participated in the creation of mankind.
- And contrary to a common belief, the Anunnaki were not the first extraterrestrials and gods to create a human from clay.
- Many other deities from different pantheons created also man from clay. For instance, Khnum "Kneph" (Meaning: To build, to unify in Egyptian) was one of the oldest Egyptian gods who created mankind from clay on a potter's wheel.
- Khnum became a variation of Ptah.
- The Anunnaki first landed in Phoenicia where they established their first colonies, and short after, they created a most elaborate medical center on the Island of Arwad, then a Phoenician territory.

- However, the Anunnaki ameliorated their genetic creations, and upgraded early human forms and primitive humans in their laboratories in Sumer.
- The Sumerian texts and their translations in western languages gave more exposure to the Anunnaki of Sumer than to the other and equally powerful Anunnaki of Phoenicia and Central Africa.
- The Sumerian texts include various versions of the creation of mankind by a multitude of Anunnaki's gods and goddesses.
- Some passages in the Sumerian texts refer to different creators, as well as to multiple genetic experiments.
- There is no reference to one singular genetic creation of the early human races, or a solid certainty to the fact that mankind was genetically created by one single god.

*** *** ***

> **A multitude of gods and goddesses created different types and categories of human beings**

To name a few:

1-Ninlil

Ninlil, also called Aruru, Ninhursag, Ninhursanga, the Lady of the Mountains, the ruler of the heavens, underworld, wind, earth, and grain, wife of Enlil and the mother of Nanna/Utu experimented with different forms and shapes of early human beings.
Ninlil also created Endiku.
In the Epic of Gilgamesh we read:
"...she created mankind...so numerous...she thrust her hands into the waters and pinched off some clay, which she dropped in the wilderness, in the wilderness she made Endiku the hero..." and in another passage, it was written: "My friend Endiku whom I loved has turned to clay...died, returned to the clay that formed him..."

*** *** ***

2-Marduk

Marduk was the son of Ea and husband of Sarpanitu, the sun-god, and also the god of war, fire, earth and heaven, and one of the major creators of heroes, gods and humans. He waged a ravaging war against Tiamat, dismembered her, and used several parts of her body to create the world and the early races of humanity.

In the Sumerian epic and scriptures, the Enuma Elish, we read:

"He opened his mouth and unto Ea he spake
That which he had conceived
in his heart he imparted unto him...
My blood will I take
and bone will I fashion
I will make man, that man may...
I will create man who shall inhabit the earth,
That the service of the gods
may be established,
and that their shrines may be built..."

3-Inanna

Inanna was the legendary Sumerian goddess who created the first 7 prototypes of mankind. Many other civilizations worshipped her under different names, such as Astarte, Istar, Ashtar, Asherat.
Inanna was called Ashtaroot and Ishtar in Phoenician; Ashtoreth and Ashtaroth in Hebrew; Ashteroth in Canaanite; Atargatis in Greek.

*** *** ***

4-Namma

The Sumerian-Anunnaki goddess Nammu "Namma" and her son Enki created multiple forms of humans, sometimes using clay, and some other times blood of warriors they slaughtered.

5-Ea

Ea killed Kingu, the demon son of Tiamat, and used his blood to create mankind.

Ea was the son of Anu. Sometimes he is mentioned as the son of Anshar. Ea created Zaltu as a complement to Ishtar. According to Assyro-Babylonian Mythology, "Ea suggested the method of creating man, in response to the heavy workload of the Igigi. Ea was a good god."

*** *** ***

Chapter 45
The Anunnaki As They Were Known to Ancient Tribes and Civilizations

- The Anunnaki as they were known to ancient civilizations...45
- The real meaning of the word Anunnaki...45
- Possibly a wrong translation of the word Anunnaki on the part of authors...45
- How the word Anunnaki was used...46
- The different meanings of the word Anunnaki...46
- The Anunnaki were known to many neighboring countries in the Near East, Middle East, by and under different names...46
- Various attributes or definitions were given to the Anunnaki...47
- The Anunnaki in the Enuma Elish...48
- Epistemology and historical terminology of the word Anunnaki...48
- Linguistic examples...49

*** *** ***

Chapter 45
The Anunnaki as they were known to ancient tribes and civilizations

The real meaning of the word Anunnaki

> Anunnaki is an Ana'kh/Sumerian/Akkadian noun that means many things.
> The Anunnaki is an ancient Sumerian/Akkadian term, that has been mistranslated and erroneously interpreted by authors and ufologists.
> It is commonly translated as "those who Anu sent from heaven to earth." This is an honest translation, however, not totally correct. The fact is that this translation is rather an interpretation than an epistemological and terminological definition.

Possibly a wrong translation of the word Anunnaki on the part of authors

Literally, it is composed of three words:
- 1-Anu (God; leader; king)
- 2-Na (To send or to follow)
- 3-Ki (Planet earth)

This is correct to a certain degree, because "Na" is not found in Akkadian or Sumerian, but exclusively in Ana'kh and Ulemite. And few scholars had access to the vocabulary of the Ana'kh.
Consequently, none of them knew, or took into consideration the meaning of "Na" in Ana'kh, because they were not aware of its existence.
Therefore, their cliché interpretation/translation should not be considered as the only and ultimate one.

*** *** ***

How the word Anunnaki was used

The word is used in a plural form.
The gods together are called Anunnaki, and are represented as ša Šamê u erSetim, meaning the Anunnaki of the sky (heaven) and planet earth. In many instances, the word Anunnaki represented two categories of gods; the Anunnaki and the Igigi.
Many contemporary writers and ufologists believe that "The Igigi in that case are the gods of heaven, while the Anunnaki refer to the gods of the netherworld, the empire of the death." Unfortunately, this interpretation is also incorrect.*

*** *** ***

The different meanings of the word Anunnaki

The Anunnaki were known to many neighboring countries in the Near East, Middle East, by and under different names

The Anunnaki were known to many neighboring countries in the Near East, Middle East, and Anatolia. Consequently, they were understood and called differently. For instance:

- **1-**The early Habiru called them Nephilim, meaning to fall down to earth.
- **2-**Some passages in the Old Testament refer to them as Elohim.
- **3-**In Ashuric (Assyrian-Chaldean), and Syriac-Aramaic, they are called Jabaariyn, meaning the mighty ones.
- **4-**In Aramaic, Chaldean and Hebrew, Gibborim mean the mighty or majestic ones.
- **5-**In literary Arabic, it is Jababira. The early Arabs called them Al Jababira; sometimes Amalika.
- **6-**The Egyptians called them Neteru.

- **7**-The early Phoenicians called them An.Na Kim, meaning the god or heaven who sent them to us or to earth.
- **8**-The early inhabitants of Arwad called then. Anuki, meaning the subjects or followers of Anu.
- **9**-The early Hyskos called them the Anuramkir and Anuramkim, meaning the people of Anu on earth. It is composed of three words: Anu + ram (People) + Ki (Earth). The primitive form of Ki was kir or kiim.
- **10**-The Ulema call them Annakh or Al Annaki, meaning those who came to earth from above.
- **11**-The Anunnaki were also called Anunnaku, and Ananaki.

*** *** ***

*Author's note: Please refer to de Lafayette's book "Thesaurus-Dictionary Of Sumerian Anunnaki Babylonian Mesopotamian Assyrian Akkadian Aramaic Hittite: World's first languages & civilizations: Terminology & relation to history, Ulema & extraterrestrials."

Considered by Assyriologists and mythologists/alternative anthropologists as a group of Akkadian and Sumerian deities, quite often, the Anunnaki were equated/associated with the Annunna, meaning the fifty great gods. Annuna was written in various forms, such as:
- **a**-A-nun-na,
- **b**-Anu-na,
- **c**-Anuma-ki-ni,
- **d**-Anu-na-ki.

*** *** ***

Various attributes or definitions were given to the Anunnaki

Various attributes or definitions were given to them, such as:

- **a**-Major gods in comparison to the Igigi who were considered minor gods.
- **b**-Those of a royal blood or ancestry.
- **c**-The royal offspring,
- **d**-The great gods of heaven and earth. An means heaven, and ki means earth.
- **e**-The messengers or subjects of god/king Anu.
- **f**-The children of Anu and Ki.

The Anunnaki in the Enuma Elish

The early Anunnaki had strong relations with the Sumerians, the Phoenicians, the Hyskos, the Philistines, the Etruscans, the Druids, the Minoans, the Atlanteans, and the people of Mu.
The Annunaki appear in the Babylonian creation myth, Enuma Elish. In the late version magnifying Marduk, after the creation of mankind, Marduk divides the Anunnaki and assigns them to their proper stations, three hundred in heaven, three hundred on earth.
In gratitude, the Annunaki built the magnificent Esagila, that rivaled Apsu.

*** *** ***

Epistemology and historical terminology of the word Anunnaki

Anunnaki is a Sumerian/Akkadian/Chaldean/Assyrian name. It is composed of:
- **1**-An=Above; sky; heaven; clouds; deity; god;
- **2**-Nak (also Nakh): From; belongs to.

To the early Phoenicians, Arwadians, Adamites and Elamites, Anunnaki meant ruling kings. But the primordial Phoenician meaning was: Those who gave us life.
This meaning/interpretation is derived for the formation of three ancient Phoenician words:
- **a**-An'kh (Life; god; spirit), so the first part "An" is used,
- **b**-Nunnak (From within),
- **c**- I.

The letter "i" is usually added to the end of a Semitic word or a name to mean one of the following:
- **A**-It belongs to;
- **B**-It came from;
- **C**-My.

Linguistic examples:

- 1-For instance, the word Ab means father in many Semitic languages (Assyrian, Sumerian, Aramaic, Arabic, etc.) When we add the letter "i" to Ab, it becomes Abi, which means MY father. The letter "i" adds a sense of belonging and origin.

Later on in history, non-Semitic people and neighboring civilizations in the Middle East, Near East, and Anatolia incorporated the letter i, in their vocabularies. However, they attached to i, the letter g. And the new addition became "gi", always meaning: Mine; my, or belongs to.

- For instance, in ancient Turkisk (Osmani language, the language of the Ottoman Empire) words like : Kahwagi or Ahwagi meant the man who makes coffe; it is composed of two words: Kahwa, Ahway, Kahwe (Coffee) + Gi.

The word baltagi is composed from two words: Balta (Ax) + Gi. It means the man with the ax, or the man who makes axes.

The word Diwangi is composed from two words: Diwan (Forum, an area or an office in a palace, a Majless) + Gi. The meaning becomes: The man who belongs to the forum, to an office of the palatial area. However, and strangely enough, sometimes, a word that is written and spelled exactly the same way in two different languages may not mean the same thing.

- For instance, Diwangi in ancient Ousmani language, means the man of the forum or an office in a royal palace. It is quite a respectable word. But in Arabic (Lebanese and Syriac Arabic), it means a charlattan or an impostor. Not so respectable, as you can see.

According to Dr. John Heise, Anunnaki is a collective name for the gods of heaven and earth, and in other contexts only for the gods of the netherworld, the empire of the death (In particular beginning in the second half of the second millennium.)

According to several linguists, the word Anunnaki is a loan word (Plural only) from Sumerian a.nun "n-a-k", meaning literarily: semen/descendants of the (Ak) monarch (Nun) and refers to the offspring of the king of heaven An/Anum.

*** *** ***

Chapter 46
Ataba-Darr-Ja and Atabukha Darja

What is Ataba-Darr-Ja?

Ataba-Darr-Ja is an Anak'h/Ulemite/Phoenician expression, composed from:
- **1**-Ataba (Door step)
- **2**-Darr (House or temple)
- **3**-Ja (Grade or level). General meaning is social classes.

The Anunnaki's society is divided into two classes: The lower class and the higher class. Both are under the control of a "Sinhar" or a "Baal-shalimroot-An'kgh."
Baal-shalimroot-An'kgh means: Greatest leader. Sinhar means: Leader or ruler.
When the word "Sinhar" is attached to "Mardack" or "Marduck", the new meaning becomes: Leader or creator of the ultimate energy.

To the Anunnaki, the universe was NOT created by God

Why "Ultimate energy" is so important? Because the Anunnaki do not believe in the God we know and worship.
To the Anunnaki, the universe was NOT created by God.
The universe is "What It Is" or "Creation by Itself".

*** *** ***

What is Atabukha Darja?

Atabukha Darja is an Ana'kh word referring to the Anunnaki's social classes:

- **1**-The lower class of the Anunnaki consists of the Nephilim.
- **2**-The higher class of the Anunnaki consists of the Sinhar-Harib.
- **3**-Baalshalimroot-An'kgh is the Anunnaki's greatest leader.
- He rules both classes.
- His name means the following:

- **A**-Baal: God; creator; the leading force of the creation;
- **B**-Shalim: Friendly greetings; message of the leader; peace; root: the way; direction of victory;
- **C**-An'kgh: Eternity; wisdom; eye of great knowledge; the infinite; the ever-lasting energy.

The second in command is Adoun Rou'h Dar, also Adon-Nefs-Beyth.

His name means the following:
- **a**-Adoun or Adon: The lord; god; The ultimate one;
- **b**-Rou'h or Nefs: The spirit; The original creative force; the soul; the mind;
- **c**-Dar, Beit or Beyth: Residence; the House of the Lord.

Members of the higher class of the Anunnaki are 9 foot tall. Their lifespan averages 350,000-400,000 years.

*** *** ***

Chapter 47
The Anunnaki view of the Afterlife

- Humans should not be afraid to die, nor fear what is going to happen to them after they die...55
- Soul is an invention of mankind. It does not exist anywhere inside us. Instead, there is a non-physical substance called "Fik'r" ...56
- What is Fik'r? ...56
- The "Fik'r" is linked to the Anunnaki our creators through a "Conduit" found in the cells of the brain...56
- We are a copy of an original copy of ourselves stored in the "Rouh-Plasma"; a compartment under the control of the Anunnaki...57
- You, your so-called "soul", your mind and your new shape in the Anunnaki's Bubble...57
- Is the afterlife a physical world? ...58
- Everything you have experienced on earth exists in other dimensions, and there are lots of them...58

*** *** ***

Chapter 47
The Anunnaki view of the Afterlife

> Although the Anunnaki do not believe in the same god we worship, revere and fear, their understanding of the creator of our universe (Our galaxy), other galaxies, the whole universe and especially life after death (The afterlife) could change the way we understand "God", the universe, the reason for our existence on earth, the concept of immortality, the fabricated organized religions we follow, and most particularily what do we expect to see, have and feel in the next wife.

Humans should not be afraid to die, nor fear what is going to happen to them after they die

Their view of the afterlife gives a great hope and an immense relief to human beings. According to the "Book of Ra-Dosh", the only Anunnaki's manuscript left on earth in the custody of the Ul'ma (Ulema), humans should not be afraid to die, nor fear what is going to happen to them after they die.

SinharMarduck, an Anunnaki leader and scholar said human life continues after death in the form of "Intelligence" stronger than any form of energy known to mankind. And because it is mental, the deceased human will never suffer again; there are no more pain, financial worries, punishment, hunger, violence or any of the anxiety, stress, poverty and serious daily concerns that create confusion and unhappiness for the human beings.

After death, the human body never leaves earth, nor comes back to life by an act of god, Jesus, or any Biblical prophet. This body is from dirt, and to dirt it shall return. That's the end of the story. Inside our body, there is not what we call "Soul".

Soul is an invention of mankind. It does not exist anywhere inside us. Instead, there is a non-physical substance called "Fik'r"

Soul is an invention of mankind. It does not exist anywhere inside us. Instead, there is a non-physical substance called "Fik'r" that makes the brain function, and it is the brain that keeps the body working, not the soul. The "Fik'r" was created by the Anunnaki at the time they created us. The "Fik'r", although it is the primordial source of life for our physical body, it is not to be considered as DNA, because DNA is a part of "Fik'r"; DNA is the physical description of our genes, a sort of a series of formulas, numbers and sequences of what there in in our body, the data and history of our genes, genetic origin, ethnicity, race, so on.

What is Fik'r?

> Consider "Fik'r" as a cosmic-sub-atomic-intellectual-extraterrestrial (Meaning non-physical, non-earthy) depot of all what it constituted, constitutes and shall continue to constitute everything about you. And it is infinitesimally small.
> But it can expand to an imaginable dimension, size and proportions. It stays alive and continues to grow after we pass away if it is still linked to the origin of its creation, in our case the Anunnaki.

The "Fik'r" is linked to the Anunnaki our creators through a "Conduit" found in the cells of the brain

The "Fik'r" is linked to the Anunnaki our creators through a "Conduit" found in the cells of the brain. For now, consider "Fik'r" as a small molecule, a bubble. After death, this bubble leaves the body.
In fact, the body dies as soon as the "bubble" leaves the body. And the body dies because the "bubble" leaves the body.
Immediately, with one tenth of one million of a second, the molecule or the "bubble" frees itself from any and everything physical, including the atmosphere, the air, and the

light...absolutely everything we can measure, and everything related to earth, including its orbit. The molecule does not go before St. Paul, St. Peter or God to stand judgement and await the decision of god -whether you have to go to heaven or hell– because there is no hell and there is no heaven the way we understand hell and heaven.

So it does not matter whether you are a Muslim, a Christian, a Jew, a Buddhist or a believer in any other religion. The molecule (Bubble) enters the original blueprint of "YOU"; meaning the first copy, the first sketch, the first formula that created "YOU".

*** *** ***

We are a copy of an original copy of ourselves stored in the "Rouh-Plasma"; a compartment under the control of the Anunnaki

Every human being has a double of himself/herself. We are a copy of an original copy of ourselves stored in the "Rouh-Plasma"; a compartment under the control of the Anunnaki on Nibiru and can be transported to another star, if Nibiru ceases to exist. And this double is immortal. In this context, human is immortal, because its double never dies. Once the molecule re-enters your original copy (WHICH IS THE ORIGINAL YOU), you come back to life with all your faculties, including your memory, but without physical, emotional and sensorial properties (the properties you had on earth), because they are not perfect.

*** *** ***

You, your so-called "soul", your mind and your new shape in the Anunnaki's Bubble

At that time, and only at that time, you will decide whether to stay in your double or go somewhere else...the universe is yours. If your past life on earth accumulated emough good deeds such as charity, generosity, compassion, forgivennes, goodness, mercy, love for animals, respect for nature, gratitude, fairness, honesty, loyalty...then your double will have all the wonderful opportunities and reasons to decide and select what shape,

format, condition you will be in, and where you will continue to live.
In other words, you will have everything, absolutely everything, and you can have any shape you want including a brand new corporal form. You will be able to visit the whole universe and live for ever, as a mind, as an indestructible presence, and also as a non-physical earthy body, but you can still re-manifest yourself in any physical body you wish to choose.
Worth mentioning that the molecule, i.e, "bubble" (So-called soul in terrestrial term) enters a mew dimension by shooting itself in space and passing through the "BAB", a sort of a celestial star-gate or entrance. If misguided, your molecule (So-called your soul) will be lost for ever in the infnity of time and space and what there is between.

*** *** ***

Is the afterlife a physical world?

Everything you have experienced on earth exists in other dimensions, and there are lots of them

No and yes.
Because life after death unites time and space and everything that it constitutes space and times. This extends to, and encompasses everything in the universe, and everything you saw, knew, felt, liked and disliked.
- Everything you have experienced on earth exists in other dimensions, and there are lots of them.
- Everything you saw on earth has its duplicate in another dimension.
- Even your past, present and future on earth has another past, another present and another future in other worlds and other dimensions.
- And if you are lucky and alert, you can create more pasts, more presents and more futures, and continue to live in new wonderful worlds and dimensions; this happens after you die. Anunnaki and some of their messengers and remnants on earth can do that. The physical aspect

of the afterlife can be recreated the way you want it by using your "Fik'r".
- Yes, you can return to earth as a visitor, and see all the shows and musicals on Broadway or hang out on Les Champs-Elysées.
- You can also talk to many people who died if you can find their double in the afterlife.
- You can also enjoy the presence of your pets (dead or alive), and continue to read a book you didn't finish while still alive on earth.
- What you currently see on earth is a replica of what there is beyond earth and beyond death.
- The afterlife is also non physical, because it has different properties, density and ways of life.

*** *** ***

Chapter 48
All That Is

To the Anunnaki, God is made of inexhaustible mental energy

The Anunnaki view of God similar to human religions in many ways, but contains much more information. The term they use to describe God is "All That Is." To the Anunnaki, God is made of inexhaustible mental energy, and contains all creation within Itself, therefore representing a gestalt of everything that has existed, exists now, or will exist in the future, and that includes all beings, all known universes, and all events and phenomena.

God's dearest wish is to share in the lives of all It's creations, learn and experience with them, but while they are imperfect, God Itself is perfect, which is why It can only be seen as a gestalt. It is possible that other "primary energy gestalts" existed before God came into being, and actually created It/Him.

The possibility exists that there are many Gods

If so, then the possibility exists that there are many Gods, all engaged in magnificent creativity within their own domains. The Anunnaki are not certain if that is so, but do not dismiss this beautiful possibility. The individuals that exist within God, though part of God, have free will and self-determination. In life and in death, each is a part of God and also a complete and separate individual that will never lose its identity.

The Anunnaki are indeed the creators of human beings

The Anunnaki are indeed the creators of human beings, but since each Anunnaki is a part of God, there is no conflict in the idea of their creation of humanity. Creation is endless and on-going, and human beings, in their turn, create as well – for example, great art, literature, and service to other people, animals, and the planet Earth – though they do not exactly create life as yet.
We are all part of the grand gestalt, and that makes All That Is such an apt name for God.

*** *** ***

Chapter 49
Sexual relationships and reproduction of the Anunnaki and extraterrestrials

- Anunnaki reproduction is done by technology, involving the light passing through the woman's body until it reaches her ovaries and fertilizes her eggs...65
- Anunnaki's sex involves an emanation of light from both participants...65
- Like many extraterrestrials, the Anunnaki do not have genital organs...66
- Major points about the subject ...66
- Aliens reproduce in laboratories...66
- Alien babies are nourished by a "light conduit." ...66

***   ***   ***

Chapter 49

Sexual relationships and reproduction of the Anunnaki and extraterrestrials

Anunnaki reproduction is done by technology, involving the light passing through the woman's body until it reaches her ovaries and fertilizes her eggs

> Sex and reproduction are two separate functions.
> Anunnaki reproduction is done by technology, involving the light passing through the woman's body until it reaches her ovaries and fertilizes her eggs. The eggs go into a tube. The woman is lying on a white table for this procedure, surrounded by female medical personnel. If performed by uncaring aliens (such as the grays and the reptilians) it is unpleasant and even can be painful, which has given rise to the abductee's stories of suffering. However, not all aliens are created equal.

Anunnaki's sex involves an emanation of light from both participants

The Anunnaki, which are a very compassionate race, are very gentle and the procedure is harmless. Apparently, the Anunnaki version of sex is much more enjoyable for both genders. It involves an emanation of light from both participants. The light mingles and the result is a joy that is at the same time physical and spiritual. The Anunnaki do not have genitals the way we do. As a hybrid becomes more and more Anunnaki, he/she loses the sexual organs and becomes physically like the Anunnaki.
The hybrid welcomes the changes and feels that he/she has gained a lot through the transformation. The Anunnaki mate for life, like ducks.

They don't even understand the concept of infidelity, and don't have a word for cheating, mistress, extramarital affairs, etc. in their language.

Like many extraterrestrials, the Anunnaki do not have genital organs

Like many extraterrestrials, the Anunnaki do not have genital organs, but a lower level of aliens who inhabit the lowest interdimensional zone and aliens-hybrids living on earth do. The stories of the abductees who claim to have had sex with Anunnaki are to be disregarded. Those stories are pure fiction.

Major points about the subject:

Aliens reproduce in laboratories

- **1**-Aliens reproduce in laboratories.
- **2**-Aliens do not practice sex at all.
- **3**-Aliens fertilize "each other" and keep the molecules (not eggs or sperms, or mixed liquids from males and females) in containers at a very specific temperature and according to well-defined fertilization and reproduction specs.
- **4**-Alien babies are retrieved from the containers after 6 months.
- **5**-The following month, the mother begins to assume her duty as a mother.
- **6**-Alien mothers do not breast-feed their babies, because they do not have breast, nor do they produce milk to feed their babies.

Alien babies are nourished by a "light conduit."

- **7**-Alien babies are nourished by a "light conduit."
- **8**-Human sperm or eggs are useless to extraterrestrials of the higher dimension.

- **9**-Extraterrestrials are extremely advanced in technology and medicine. Consequently, they do NOT need any part, organ, liquid or cell from the human body to create their own babies.
- **10**--However, there are aliens who live in lower dimensions and zones who did operate on abductees for other reasons – multiple reasons and purposes – some are genetic, others pure experimental.

*** *** ***

Chapter 50
Immaculate Conception:
A Typical Anunnaki Pattern

- Cherchez the Bible! ...71
- Impregnated women by "something" or "somebody" other than their husbands...71
- Got pregnant before they were married...71
- Almost every famous and notable female in the Bible had some sort of a mind-boggling pregnancy situation...72
- Here are some of the famous women of the Bible who got pregnant by something or somebody else than their husband...73
- Pregnant before marriage...73
- Very old and barren women...73
- Desperately old women...73
- Barren until middle age...73
- Famously barren...73
- Was David an Anunnaki offspring? ...74
- In the Bible, so many of the important female characters are barren and unable to conceive – until a very special person comes, sometimes as an angel, to help...75

*** *** ***

Chapter 50
Immaculate Conception; a typical Anunnaki pattern

Cherchez the Bible!

> The Anunnaki live for thousands of years, and their understanding of history is very deep. In fact, the Anunnaki dictated and created humanity's history. We, who live such short lives, make many historical mistakes, even when written records are available. Some of the most ironic, and enigmatic aspects of these records are found in the most revered, sacred, read, bought, sold, and fanatically preserved depot of human history, soul salvation and consciousness: The Bible!

Impregnated women by "something" or "somebody" other than their husbands

Impregnated women by "something" or "somebody" other than their husbands, and how all these pregnancy mysteries and extramarital affairs are related to the Anunnaki. Here, we have two intriguing situations:

- **A-Got pregnant before they were married**: Were these Jewish and Christian Women of the Bible Impregnated by the Judeo-Christian God, The Holy Spirit, the Angels, the Anunnaki or some unknown Lovers?
- **B-They are married now:** They admitted that they had no intercourse with their husbands, and that *the* husband was Not the father of the child...some were 70 years old, others barren, yet they got pregnant!

Who did it?
And how did they get pregnant?
Did the Angel of God come back to do the same thing, his regular business?
Would it be easier to assume that these women had an extramarital affair?
Perhaps, they had an intercourse with an alien as many women nowadays claim?
Who is that "thing", "that invisible power", or that" mysterious incognito" who did it?
- An angel?
- A spirit?
- A secret lover?
- A rapist?
- An alien?

Or just perhaps a husband who did it and simply forgot about it?
The Bible is full of stories like these and steams with sex! If you still think that the resurrection of Jesus Christ is the greatest mystery of all time, and the most amazing and dazzling event in the Bible? Think again!
This is not about conjugal infidelity or cheating on your spouse, or having an extramarital affair. On the surface that is!

*** *** ***

Almost every famous and notable female in the Bible had some sort of a mind-boggling pregnancy situation

Almost every famous and notable female in the Bible had some sort of a mind-boggling pregnancy situation, a fairy tale genre. Here is the scenario:
- **a**-They shared one extraordinary thing in common: All conceived without having an intercourse with their husbands;
- **b**-Some delivered a baby before they were married;
- **c**-Many were either impregnated by another man, another something, a divine spirit, a messenger from god, but never by their husbands, fiancés or the man to

be wed to! while Others were very old to give birth to a baby;
- **d**-And the rest were certified barren.

These pregnancy situations are taken directly from the Bible. They are Bible certified.

*** *** ***

Here are some of the famous women of the Bible who got pregnant by something or somebody else than their husband

Here are some of the famous women of the Bible who got pregnant by something or somebody else than their husband...some were barren...others very old.

A-Pregnant Before Marriage:
Virgin Mary gave birth to a child, Jesus, before she was married or even knew her old husband
B-Very Old and Barren Women:
Elizabeth (Elisheva) wife of the very old Zecharaia, and aunt of Jesus Christ, gave birth to John the Baptist at 55 or 60.
C-Desperately Old Women:
Sarah conceives at a ridiculously old age.
D-Barren until Middle Age:
Rebekah was married to Isaac and was of a middle age when she conceived. And this happened only when Isaac has prayed to God.
E-Famously Barren:
Hannah was barren and married to a very old man.

*** *** ***

One of the most fascinating points of these stories is that two women in the Bible gave birth to Anunnaki hybrids, and the Doctors of the Church called the pregnancy the fruit of an immaculate conception.
Check out this:
- **A-**While Batya did not need to marry, Miriam did, and her husband, was exactly the type of husband Mary of Nazareth married.

- **B**-Think of Hannah. She is barren. She goes with her kind, older husband to the temple, and prays for a son. In the Bible, a priest named Eli comes to her and promises her a son, if she will be willing to dedicate the boy to God. This does not hurt the boy in any way, on the contrary, little Samuel gets a very good education out of the bargain, but of course the priest was never a priest. He was an Anunnaki, and he helped the barren woman get pregnant in his scientific way.

And what was the destiny of this child?
He was the one who would eventually anoint young David and make him replace the clinically depressed, homicidal, suicidal King Saul...
And who is David, then? Young David is the son of Jesse, son of Obed, son of Boaz. And Boaz is the husband of Ruth. Yes, the famous Moabite lady who lost her husband to a plague and moved away from Moab with her beloved mother-in-law, Naomi, to the Land Judah. What the Bible does not tell us is that the beautiful young widow, whose genetic makeup was just right, was contacted by an Anunnaki on the way to the Land of Judah, and with Naomi's consent, accepted the Immaculate Conception very gratefully.

Was David an Anunnaki offspring?

Then, when settled in the Land of Israel and again with Naomi's consent, she meets Boaz, a distant cousin, a kindly, older man of the Joseph and Jethro type, who wants to marry her and give a good home to the two women and their little Anunnaki hybrid boy. Incidentally, Ruth lived to an advanced old age. So old, that she knew her great-grandson, David, and could smile at his incredibly good looks, complete with red hair, blue eyes, and glowing olive skin, which he had inherited from her Moabite side. For once, her genetics won over the usually occurring black eyes and hair of the Anunnaki, and we know what a lady-killer David turned out to be. And they are not the only ones.

*** *** ***

In the Bible, many important female characters are barren and unable to conceive – until a very special person comes, sometimes as an angel, to help...

A-Sarah conceives at a ridiculously old age.
B-Rebekah needs her husband's Isaac special prayer to conceive – and probably a lot more.
C-Nothing can be more intriguing than the story of Elizabeth (Elisheva in Hebrew, meaning My God is an oath) who was the cousin of Mary of Nazareth. There are various versions about her life, but amazingly, much of these stories are absolutely true. Yes, she was the cousin of Mary, daughter of a sister of Mary's mother.
She was married to Zecharaia, a good man who was typical to the Anunnaki-human hybrid's stepfather, much like Jethro and Joseph. Elizabeth reached an advanced age (though she was not as old as seventy as some sources claim) without conceiving, when an angel, which of course was an Anunnaki, visited the couple and promised them a son. The scientific Anunnaki system soon impregnated Elizabeth, and the son turned out to be John the Baptist, another hybrid with magical qualities.
Elizabeth was chosen for the task for her perfect genetic makeup. She was a direct descendant of Aaron, Miriam's son.

The intricate, thousands of years long records of such relationships, are meticulously kept by the Anunnaki. And so the pattern returns, again and again, until it culminates in an entire religion, Christianity, based on an exceptional little boy named Yeshua (Yashou, Yassou'h, Jeshua, Jesus) who is also a hybrid, born of an immaculate conception, and of an unknown father!
And as long as the connection between Anunnaki and humans continue, the pattern will go on.

*** *** ***

Chapter 51
Al-Jahiliya

- What is Al Jahiliya? ...79
- But how does this fit in the Anunnaki scenario? ...79
- The Raheb Bouhayra, Prophet Mohammad, Islam and Anunnaki connections...80
- The practice of the occult and the study of extraterrestrial literature were abolished by Islam...81
- The society of Ulema of Rama-Dosh...81

*** *** ***

Chapter 51
Al-Jahiliya

What is Al Jahiliya?

> Al Jahiliya is a key word in the history of the Arabs and more precisely in Islam. Jahiliya means the "Years of ignorance", and the "Years of darkness." It was a word created by the early followers and companions of the Prophet Mohammad to refer to the years and ages before the birth of the Prophet, and the arrival of Islam.
> These years and ages were considered and called the years of ignorance because they preceeded the birth of Mohammad.

With his birth and the message of Islam, the light shined on humanity, and everything that existed before him was darkness. And all those who lived in those years were called "Al Jouhal", which is the plural of "Jahel" meaning "Ignorant"; ignorant of the truth, the salvation of humanity and everything between and beyond.

*** *** ***

But how does this fit in the Anunnaki scenario?

During the so-called years of "Jahiliya", many thinkers, philosophers, writers and poets believed in the Gnostic teachings, the early form of Christianity, and the Anunnaki-Ulema doctrine. At that time in history, Islam was not born yet, and the Arabs who lived in that part of the world (Middle East, Near East, North Africa, the Arab Peninsula, etc...) were either pagans, Christians, Jews or believers in other religions.
Amid these avalanches of religions, metaphysical and spiritual concepts and beliefs, many Jews and Christians believed in

different and contradictory Hebrew and Christian dogma and theologies.
This included the concept of the Messiah, the resurrection of Jesus Christ, the divinity of Jesus, the origin of man, and of course the Anunnaki's genetic creation of the human race.

Some of these avant-garde thinkers in that part of the world, were Arabs who belonged to Gnostic and extraterrestrial entourages.
Worth mentioning here, that non-Arab, and non-Muslim men of science and letters also believed in the Anunnaki's connection. The most illustrious ones were the Ulema.
They were familiar with the Sumerian deities, accounts of super-beings from a different world, the "high spirits", the "Unseen entities", the "celestial human-angels", and the Anunnaki.

The Raheb Bouhayra, Prophet Mohammad, Islam and Anunnaki Connections

The secret "Book of Ra-Dosh", the official bible of the civilizations and teachings of the Anunnaki was in the hands of eminent Arab thinkers and poets.
Accounts of the era tell about how the Prophet Mohammad learned by heart the verses of the Quran from a Christian ascetic Ulema, called "Raheb Bouhayra" who lived in the desert. This monk was one of the messengers of the Anunnaki in the Arab Peninsula.
He was a poet, a visionary and a holy man. With Islam invading the Arab world, and the Prophet winning battles one after another, many Arabs converted to Islam, either by conviction or by the power of the sword.

Same thing happened in Christianity history with the rise of the Templar Order, and the hysterically fanatic Spanish Conquistadors, no difference whatsoever!!
Organized religions are poison!

*** *** ***

The practice of the occult and the study of extraterrestrial literature were abolished by Islam

With the sharp rise of Islam, any interest in anything non-Islamic disintegrated.
Nothing was bigger, more powerful, more important, and more authoritative than Islam.
The Spiritists known as the Rouhaniyyin and/or Al-Mounawirin who were the early custodians of the Anunnaki's extraterrestrial scriptures and literature vanished from the face of the "Islamic Earth".
Consequently, talking about Anunnaki and extraterrestrials became blasphemous.
Bear in mind, that even the greatest Arab minds, scientists, historians, physicians, and poets of the era who were born before the birth of Mohammad and the arrival of Islam were considered ignorant, and even sinners by Islamic standards. The Prophet himself called them "Min Ahl Al Nar".
Verbatim itmeans: From the the people of fire. He meant by that: They are from hell. In other words, they are sinners and they are going to Jahanam "Hell".

*** *** ***

The Society of Ulema of Rama-Dosh

However, a group of enlightened Arab teachers (Christians, Jews, liberal thinkers, and free-spirited philosophers) created a secret society called the "Ulema of Rama-Dosh". They were closely related to the freemasons in Lebanon, Cairo, Damascus, Island of Awad and the remnants of the Templar Order in Cyprus, and the St. John Order of Malta. Thanks to the Ulema of Rama-Dosh, Anunnaki oral history, esoteric practices, and secret manuscripts have survived.
Later on in history, the Arabic pre-Islamic word Ulema was replaced by the Arabic Islamic word Allamah or Al Hallama. However, the Ulema and the Allamah were very different from each other in many ways.

The Ulema remained the custodians of the Anunnaki's secret knowledge and esoteric powers, while the Allamah were considered as the "Alamin", the learned ones and leading Islamic figures of letters, literature, science and religion. Nevertheless, many Muslim teachers and spiritualists remain Ulema at heart.

Many of them - secretly of course - joined the circle of the non-Muslim Ulema to learn the ultimate knowledge acquired from non-terrestrial beings. The Suphists were the first to join the Non-Muslim Ulema.

Worth mentioning, that around, 850 A.D., Ulema and Allamah were semantically overlapping each other. And both words came to mean or express the same thing in the Islamic and Arab world. Many Soufiyyin (Sufis) by joining the Ulema learned some of the greatest secrets of the Anunnaki, such as the techniques of Tay Al-Ard, Tay Al Makan, mind transmission, telepathy, and physical dematerialization.

*** *** ***

Chapter 52
The Anunnaki's Ra-Nif

What is Ra-Nif?
The many different forms of spirits

The Anunnaki referred to many different forms, shapes and "rating" of entities known to the human race as "Spirits" and "Souls".
"Ra" and "Ra-Nif" from the Anakh (Anunnaki language), gave us the words "Rouh", "Rafesh", "Nefes", and "Nefs" (in Arabic, Aramaic and Hebrew.)
In Islamic literature, these Spirits were called "Arwah Atilah" (Bad Spirit, Evil Spirits), "Afrit", "Jin", and even "Chitan" or "Sheetan". It is "Shaytan" in Aramaic and in several ancient Semitic-Acadian languages; all having the same meaning: Shaytan means the Devil or Satan.
They were also mentioned in the Bible.

*** *** ***

The union between the Anakim or Anaki and the daughters of men (Women of the earth) created a new race of beasts and demons

The union between the Anakim or Anaki and the daughters of men (Women of the earth) created a new race of beasts and demons.
This new race was the first quasi-human-animal breed on earth. Sometimes, these spirits appear to humans in the form of a "Dijjal", an Arabic word meaning "Impostor".
Here are some excerpts from the Book of Enoch: Chapter 19;...And Uriel said to me: "Here shall stand the angels who have connected themselves with women, and their Spirits assuming many different forms are defiling mankind and shall lead them astray into sacrificing to demons as Gods, here shall they stand

till the day of the great judgment in which they shall be judged till they are made an end of...And the women also of the angels who went astray shall become sirens."

On the subject, the Ulema stated that humans' desire of the flesh, Anunnaki new discovery of the physical pleasure and their intercourse with earth-made women deteriorate the spirituality (Mental faculty) and ethics of the Anunnaki living on earth. However, The Book of Rama-Dosh explained that the mating of humans and the Anaki, Ana-kim, Anunnaki or Anakh, produced most unusual hybrid beings.
The Book did not describe them as humans or Anunnaki. Instead the name "Jins" meaning breed, race or creatures was given.

*** *** ***

Chapter 53
Kefitzat ha-Derekh

- What is kefitzat ha-derekh? ...87
- The Talmud, however, mentions three Biblical individuals who experienced it...87
- Kefitzat ha-derekh can happen spontaneously, as a miracle performed for the benefit of a just and good person who is in trouble...87
- Such a miracle is assumed to be performed either by God himself, by one of his angels, or by Elijah the Prophet...88
- The secret knowledge of the holy names...88
- Baal shem and Baal Shem Tov...89
- Kefitzat ha-Derekh is only one of the many wonders the Baalei Shem...89
- The concept appeared in many areas of the world...90
- The idea resembles teleportation in other world myths and legends...91

*** *** ***

Chapter 53
Kefitzat ha-Derekh

What is kefitzat ha-derekh?

> From a joint writing project with Dr. Ilil Arbel
> Translated literally, Kefitzat ha-derekh means "the jumping of the road."
> It is usually interpreted, however, as "the shortening of the way."
> The phenomenon consists of the swift arrival of a person or persons to a distant destination, accomplished by supernatural means.
> The travelers must break the laws of nature to fit the concept, and the distance cannot be covered as quickly by walking or riding an ordinary horse, mule, or donkey. Kefitzat ha-derekh does not appear in the Bible.

The Talmud, however, mentions three Biblical individuals who experienced it

The Talmud, however, mentions three Biblical individuals who experienced it. The actual term used in the Talmud is slightly different, though. It appears as "those for whom the earth jumped (*kefitzat ha-aretz*)."
These were:
- 1-Eliezer, the servant of Abraham,
- 2-Jacob the Patriarch,
- 3-Abishai ben Zeruiah.

Kefitzat ha-derekh can happen spontaneously, as a miracle performed for the benefit of a just and good person who is in trouble

There had been some debates weather it is exactly the same concept as Kefitzat ha-derekh, but most scholars agree that it is close enough to be considered so.

In fact, Kefitzat ha-derekh can happen spontaneously, as a miracle performed for the benefit of a just and good person who is in trouble. It is usually a man; I have never encountered a story involving a woman who had Kefitzat ha-derekh. The man may be away from home before the beginning of the Sabbath, or unable to reach a place where he had promised to perform a valuable religious service.

Suddenly, he finds himself in that distant spot, sometimes without realizing how it happened, sometimes by being transported through the air or over water.

*** *** ***

Such a miracle is assumed to be performed either by God himself, by one of his angels, or by Elijah the Prophet

> Such a miracle is assumed to be performed either by God himself, by one of his angels, or by Elijah the Prophet.
> The other approach to kefitzat ha-derekh was accomplished deliberately by a group of people called "Baalei Shem.
> "The term means "masters of the Name" and the word "baalei" is the plural of "baal," or master. These people performed what amounts to magic, despite the fact that Judaism had always objected to any form of it; the Bible even recommends killing all witches.

The secret knowledge of the holy names

But this did not stop the practitioners of practical Kabbalah from being wonder makers. The baalei shem maintained that they had secret knowledge of the holy names, and that they could achieve supernatural results using them. The Name, holy Name, or Shem Tov (good Name) may be one of the divine names, the name of an angel, or a combination of letters in those Names. Most people familiar with Judaism know the name of the Baal Shem Tov, the founder of the Hasidism.

His real name was Rabbi Israel ben Eliezer. A truly great scholar, he created a new philosophy, functioned as a religious leader, and performed miracles as a wonder-maker.

Most of what we know about him is second hand, stories told by his disciples and later repeated for generations, much like Rabbi Hillel, Buddha, Socrates, and Jesus. The most famous book about him is Shivhei ha-Besht (Besht is the Hebrew acronym for Baal Shem Tov), a collection of stories that have been used in every book written about him. However, many people do not know that he was not the first to be the master of Names. Many of the stories in the book were borrowed from original tales about other "baalei shem" that had preceded the Baal Shem Tov.

Baal Shem and Baal Shem Tov

Scholars, particularly Gershom Scholem, proved that there was no difference between the words "baal shem" which means, "master of the name" and "Baal Shem Tov," which means "master of the good name." All names were good -- the baalei shem would not use them otherwise. They never performed anything even remotely negative like sorcery, black magic, or Satanism; the entire purpose of the wonders they performed was positive, and based on deep faith in traditional Judaism. The names they used could be either spoken, or written on amulets made of paper or parchment. Baalei shem are not mentioned in the Bible. They appear for the first time in the post-Talmudic period in Babylonia, or possibly at the beginning of the Geonim period, and the tales developed into the Middle Ages. The 16th and 17th centuries are extremely rich in stories, in both Israel and Europe.

*** *** ***

Kefitzat ha-Derekh is only one of the many wonders the Baalei Shem

Kefitzat ha-Derekh is only one of the many wonders the "Baalei Shem" performed.

- **1**-They could preserve the bodies of the dead as "dead-alive" by placing written amulets in the bodies, to keep them for burial in the proper time and place.
- **2**-They could create golems exorcize demons and dybbuks protect people against their enemies on both land and sea, summon beasts from the spiritual realms, send and interpret dreams, and raise the spirits of the dead.

Every process had its own formula and name, and those of kefitzat ha-derekh were different from all the others. The first known text to mention kefitzat ha-derekh, coupled with the personality of a baal shem, was a question sent by a North African community to Rabbi Hai Gaon. It described how a famous baal shem was seen in one place on the Eve on Sabbath (Friday). Later on the same Friday night, he was seen in another place, a distance of a few days journey. On Saturday evening, he was again seen back in the original place.

There was no logical way to interpret the sightings, and the community wanted Rabbi Hai Gaon to explain the miracle. In his response, Rabbi Hai Gaon categorically denied the possibility of kefitzat ha-derekh; most rational rabbis did not want anything to do with these fanciful ideas. However, the population, greatly encouraged by the baalei shem, did believe in kefitzat ha-derekh.

*** *** ***

The concept appeared in many areas of the world

The concept appeared in many areas of the world.
- **A**-Southern Italy produced a particularly famous manuscript; Megillat Ahimaaz (also called Megillat Yuhasin). In this tale a most amusing use of the formula is described -- the name was written on the hooves of the horse carrying the baal shem!
- **B**-There are tales from Germany, Poland, Russia, Spain, and others. Many more are attributed to the Ari (Rabbi Isaac Luria), the great kabbalist from Safed, Israel, and to his student, Rabbi Hayim Vital.

These wonderful tales continued to develop until the traveling rabbis abandoned walking and riding in favor of technological progress.
They started using fast ways of transportation, such as trains, to reach important destinations. As miraculous teleportation was no longer urgently needed, the telling of Kefitzat ha-derekh tales dwindled and eventually stopped.

*** *** ***

The idea resembles teleportation in other world myths and legends

The demise of this myth is surprising, since there are such things as derailed trains, car accidents, and delayed planes.
Why not have a miracle in a crowded airport, or while stranded on a lonely road in a stalled car?
The stories might have continued to accommodate such issues, and their organic growth into modernity would have been of interest. However, perhaps they did not entirely vanish, after all. The idea resembles teleportation in other world myths and legends, not to mention science fiction and fantasy, which freely make use of it in books, movies, and television. If Kefitzat ha-derekh sounds familiar to readers of science fiction, it is because Frank Herbert used this term in his book "Dune", where the concept charmingly, if somewhat inaccurately, refers to a person whose being represents the shortening of time leading to a certain important future event.

But the most famous form of modern science fiction/Kefitzat ha-derekh will be familiar to just about everyone: "Beam me up, Scotty!" Noted in the joint work of Dr. Ilil Arbel, and Lafayette.

*** *** ***

Chapter 54
On the Road to Ultimate Knowledge

- Excerpts from the book "On the Road to Ultimate Knowledge: Extraterrestrial Tao of the Anunnaki and Ulema...95
- Meeting the master, and mastering fears...95
- Scene: The little Germain is walking with his mother in the streets of Paris...95
- Year: 1945, right at the end of World War Two. France has been liberated by the Americans...95
- A group of six or seven, wearing black clothes and berets and carrying large, very visible guns, were dragging a young girl...95
- All during the war my room was the only place where I felt entirely safe...97
- Meeting my future Master...99
- In the living room stood a person that seemed to me like a materialization of a cartoon...101
- The first miracle...103
- Leaving Paris...104
- Arriving to Benares...105
- My first encounter with the paranormal...105
- And the snake obeys the Master...105
- In the house of the Master...106
- Temples, flowers, and lit oil lamps, to float on the river...106
- The magical room of colored papers...107
- And then came the miracle of the tree...108
- Miracles and tricks of illusion in the streets of India...109
- Return to Paris...112
- On the road again...on my way to Kowloon...113
- The mirage island...114
- The small temple on the top of a hill...115

- Talking to the Buddha...115
- The villagers, poor as they were and experiencing constant struggle for survival, were unusually loving, and happy...117
- Feeding the fish with food from heaven...118
- The secret is how to use your hands properly...119
- The Master: "You will find that animals understand more than people do, if you love them." ...120
- Ballet with nature...121

*** *** ***

Chapter 54

Excerpts from the book "On the Road to Ultimate Knowledge: Extraterrestrial Tao of the Anunnaki and Ulema

Co-authored by Maximillien de Lafayette and Ilil Arbel

Meeting the Master, and Mastering Fears

> This new and exciting book reveals the way to the ultimate knowledge through the life journey of a remarkable man who had studied with the greatest Ulema masters, themselves students of the Anunnaki. From early childhood he had traveled all over the world, witnessing events and learning secrets that are known to only the very few.
> The story is full of entertaining and amazing adventures, but this is not all it offers. Spending some time studying the techniques presented in it is an unparalleled opportunity to enrich your mind and your world, open yourself to miracles, and experience the universe in a unique way.

Scene: The little Germain is walking with his mother in the streets of Paris.
Year: 1945, right at the end of World War Two. France has been liberated by the Americans.

*** *** ***

A group of six or seven, wearing black clothes and berets and carrying large, very visible guns, were dragging a young girl

"Maman, what are they doing to this girl?" I screamed, trying to hide my face in her sleeve so as not to witness the horrible spectacle. "Why isn't anyone helping? Help her, Maman!"

A group of six or seven, wearing black clothes and berets and carrying large, very visible guns, were dragging a young girl, who was struggling and crying for help, to a makeshift station made of a rickety table and chair. They brutally forced the girl to her knees, and someone, holding a pair of large scissors, started cutting off her hair, pulling it mercilessly in the process. People went on walking, ignoring the horror, while some other stopped to watch, enjoying the cruel spectacle.

My mother sighed deeply. "I'll explain later," she said, and gently giving my trembling hand to my nurse, stepped forward toward the gang. "Stop this immediately!" she commanded. "Now! Release this girl at once!"

The leader of the gang turned, tremendous amazement registering on his face. He obviously did not expect anyone to approach him, let alone a woman. "And who, in the name of the Devil, are you to stop us?" He said. "You know perfectly well that we are members of the Resistance, and she is a filthy Collaborator, a friend of the Germans! We will catch all these sluts and shave their heads! This is our revenge and don't you interfere!"

"A Collaborator," my mother said contemptuously. "Half the country were collaborators. The government itself collaborated. Are you pursuing the powerful people who sold us to the Germans? No, you torment helpless little girls, who did nothing more than trying to survive. Brutes!"

Profound silence spread over the scene. The leader hesitated, not knowing exactly how to react. But he realized that if he gave way, his supremacy over his gang would end there and then. Mastering his courage, he stepped toward my mother and grabbed her hair, which, as usual, was put up in a neat twist.

It came off and her glorious golden curls fell down to her waist. "Perhaps you have done the same thing, Madame?" he sneered. "Maybe we should give you the same treatment and cut off your pretty hair? Did you play nicely with the Germans?"

I could not see Maman's face, since her back was turned toward me, but I could imagine how angry she looked; she could be quite intimidating.

As I was struggling to free my hand from my nurse, who was holding it tightly, and run to Mama, I saw her, in what seemed like slow motion, straightening herself to her commanding height, raising her hand, and slapping the leader's face with all her might. He recoiled, shocked.

"Do you know who I am?" she said in a voice that could only be described as low and menacing. "I am the widow of Charles Lumière."

"My God," said the leader, his face turning pale. "My God. Madame Lumière, forgive me. Please. I did not know." He turned around and said to the men who were holding the girl, her hair already partially cut. "Let her go. This is Madame Lumière." They instantly obeyed and the girl fell on the pavement.

"Come along, my little one," Maman said to the girl and held out her hand, "I will help you." The gang stood around them, silent, as the girl pulled herself up, helped by Mama.

"Don't ever show your faces in this street," said Maman to the leader. "You are a bad influence on my son, who will not forget this day when he grows up. Go now."

Of course I did not know it at the time, but few people's names were more respected by the Resistance than that of my father; he had done more for the Resistance than I care to explain now, and for France in general. The gang turned and left, and we went home, the girl, weak with terror and fatigue, supported by my nurse.

"Were you really a Collaborator, my child?" asked Maman, stroking the poor girl's tear-stained face.

"Yes, Madame," said the girl candidly. "But you see, father was killed in the war, my mother was sick, and my little brother was hungry... I am ashamed of myself for what I did, but yes, I had to feed him or he would have died... and then Maman would have died of a broken heart... and it was just a few times, really, Madame..."

"Don't think about it, my dear," said Maman. "We all have to survive, one way or another. I will help you and your family leave Paris. I will find you a place and some work in the country, where I have property and connections, and where you can have a peaceful life, and I promise to keep an eye on you in the future. But first of all, we must find you a wig so no one can tell what these thugs did to you."

All during the war my room was the only place where I felt entirely safe

Our house, an elegant and luxurious three-storied mansion that fortunately suffered little damage during the war, was only a minute away, and with considerable relief I entered and ran up the marble staircase to my room.

All during the war my room was the only place where I felt entirely safe, for some reason. Perhaps the pleasant presence of the familiar toys, perhaps the bed, where I could bury my head in the pillows, exercised calming influence, I really don't know.
My mother, who was invariably kind and protective, tried to have me sleep with her when I was wrecked by nightmares, but while I wanted her near me, I still needed the comforting surrounding of my room, like a cocoon, around me.

Even the portraits of our ancestors that were hanging in various rooms and on the staircase frightened me when night fell, though during the day they did not worry me. Sylvie, my little sister, had no such terrors. A very placid child, she was not traumatized at all, thank goodness, and her cute little face was always smiling.

On that afternoon I threw myself on the bed and started crying uncontrollably. A year before these events my father was killed in the war. Now, even though the war just ended, I was completely traumatized by the experience and by his death, since I was very close to him and loved him intensely. Only six years old, I could not remember a time of peace. I was terrified by the Germans who had occupied Paris, and I did not believe that the bombs will not come back. I had horrible nightmares, every single night, and woke up screaming and shaking. And now, seeing our own people abuse a helpless girl was just more than I could bear.

After a few minutes, Maman came in and took me in her arms. I was shaking with sobbing. "I want to go away too, Maman. If you can send this girl away to the country, why can't we all go? It is hateful here. There are bombs, and fires, and people attack girls in the street and I am so scared. And the man attacked you, too, he pulled your hair, Maman, I saw it. They may attack my aunt, or Sylvie, too. Why don't we leave here?"

"But I won the struggle, Germain. This thug was afraid of me, and left me alone and slunk away, the coward. Anyway, I can't leave Paris. I have to attend to the business that Papa left to us. So many people depend on it, my dear, and Papa would not have wanted me to desert them. Some day, when all is back to normal, we may travel. But in the meantime, I may find a way to give you a nice long vacation away from Paris, if you like."

"But will you and Sylvie be safe without me taking care of you, Maman?" I asked, in all seriousness believing that in some way I protected my mother and my little sister. Mama did not laugh, but answered quite seriously and reassuringly.

"Oh, yes, my dear, we'll be quite safe. Your aunt, and the servants, and our friends, and all the employees in the business, will all be here for me. And as you realize, I do know some very important people who won't let me get in any trouble, ever. You should not worry. You need to be away from Paris for a little while, just to relax and stop your bad dreams, you see? We will find a way, and it will be fun for you. And Paris is liberated, Germain. There will be no more bombs, no Germans, it will be peaceful from now on."

I felt better. I still did not want to go out of the house, not that day, anyway, and I was not entirely certain that Maman was right about the war really ending. Even though she was usually right, she still might be mistaken, the Germans might have tricked her to believe that... Nevertheless, I felt I could leave my room. I stepped over with Maman to Sylvie's room and we played with the train set and I forgot the horrors for a little while.

*** *** ***

Meeting my future Master

A few days later I came down the staircase, and was about to enter the formal living room when I was stopped by a voice I did not recognize. While I knew perfectly well that listening at doors was not a polite thing to do, I decided that hearing if what was going on in the living room was safe for me to encounter was more important than good manners.

"Well, Madame Lumière, I will be leaving tomorrow. I have met all the people I needed to meet. But I am coming back in three months, for the second set of meetings." The voice spoke perfect French, but with a foreign accent.

"Did you accomplish everything you wanted to do, Your Excellency?"

"I am not sure. It's not easy to gauge. Since Indochina is still a French territory, we cannot measure my situation by political standards. There was a need for new connections between our leaders, and at this time the authorities back home thought that a scholar would be more appropriate than a government official for such discussions. But did I succeed in all I wanted to accomplish? Who is to know? I have done my best."

"My husband would have liked this approach," said Mama. "He always believed that the scholars and thinkers could help the world much better than the political figures, who are usually only out for personal power."

"Yes indeed, Madame, your husband and I always felt the same about such matters... I will never stop missing him. Such a good friend he was to me."

"My son is devastated by his death, Your Excellency. Traumatized, to be honest," said Maman. "Your suggestion of taking him with you is more timely than you could possibly imagine."

"I believe that being away from here for three months, in different countries, different cultures, will be highly therapeutic," said the voice.

"Where do you plan to go?"

"I expected to go to Hong Kong and Indochina, but now it seems I will have to spend some time in Benares, where I currently live with my family. That is why it would be so nice for the child.

We have a big house with a huge garden, a large extended family, and little boys his own age coming to my school. The cheerful atmosphere will give him great relief and amusement. And I will be back with him in early December."

"Sounds just right," said Maman. "As I am sure you know, the school year in Paris has been postponed to January next year, since they have to allow evacuees to come back. Also, so many people are out of the country altogether. I can register him in December, and it is perfect timing for his little vacation. But what is the weather like in Benares? I am ashamed to admit, but I do not remember when the monsoons strike the area."

"It's quite safe. The monsoons strike from June to August, and they are over now. I would not expose a Parisian boy to the illnesses that might be triggered by the extreme humidity of the monsoon season. The weather is quite comfortable now, sunny and pleasant."

"Well, I can see no objection, and I am very grateful for the offer," said Maman.

"Can I see little Charles?" said the voice. It was customary in those days to call a child by the name of his father, adding the word "little" before it.

"Of course," said Maman. "Let's see how he feels about the idea. I will call him." Realizing that in a minute Mama will be out and catch me eavesdropping, I quickly dashed upstairs and sat on the bed in my room, looking as innocent as I could, holding a picture book. Presently Maman came in and told me that there is a gentleman, a friend of Papa, who wanted to meet me.

*** *** ***

In the living room stood a person that seemed to me like a materialization of a cartoon

In the living room stood a person that seemed to me like a materialization of a cartoon. He was tall and extremely skinny, had a long, white, thin beard, and light, golden skin. He wore foreign clothes that I have never seen before. But the strangest thing was the light around his head.

As he stood against the burgundy curtains that covered the window, the light was shining like a halo. I did not understand it, and Maman made no comment about it so I was not even sure if she noticed it.

Later, the Master explained to me that he put it on for me on that occasion because it had the capacity of calming me down. I don't remember seeing it around his head again. His face had a look of benevolence, deep kindness that cannot be described.

In Western culture we would refer to it, perhaps, as a biblical or saintly expression, but the Master would have never accepted such a term.

He was, anyhow, extremely appealing to everyone, so much so that I noticed later, when the servants brought tea, that they hung about the room, not wishing to leave. They seemed to be mesmerized by him. At the time, the Master was fifty-six years old, but he seemed much older in one way, and ageless in another.

"Hello, Germain," said the apparition.

"Hello, Monsieur," I said politely.

"You should say 'Your Excellency,' Germain," said Maman.

I considered that. No, he did not look like that; I saw many ambassadors in our house. He looked more like a teacher. I said that and the Master laughed. "Indeed, perhaps one day I will be your teacher," he said. "Why not call me Master, like my other students?" Yes, I thought. That fitted him very well. "I like that," I said. "Master."

"I was a friend of your father," said the Master. "And now I am a friend of your mother, and I hope your friend too. I come from Indochina."

"I see," I said noncommittally. I had no idea where Indochina was.

"How would you like to come with me on a long vacation?" he asked. "I can show you interesting foreign countries, and you will meet a lot of nice people and see strange places."

I have already made up my mind that I would go. I dearly wanted to get away from Paris, and it was clear to me that Mama thought it was a good idea.

And somehow, the Master had a strong appeal to me. But it was important to pretend that I knew nothing about the plan. I suspect the Master knew all along that I was eavesdropping before, but he did not say a word about it, then or afterwards.

"How long will the vacation be?" I asked.

"Until September, when you go back to school," said the Master.

"Okay," I said. "I will go."

"Say thank you," said Mamann, always anxious about my proper upbringing. The Master laughed again, in a most good-natured way.

"Thank you," I said. "When are we leaving?"

"How about tomorrow?" said the Master.

"Good," I said.

"We don't need too many things," said the Master to Mama. "In Benares, he will wear Indian clothing, which are cool and comfortable for the climate. With so many children around, we are always shopping for clothes and other things for them, and we can outfit him very nicely."

*** *** ***

The first miracle

The plan seemed very reasonable to me, and I settled to have some tea. After tea, when the Master prepared to leave, Maman asked him for a favour.

"I have a woman staying with me, a nun," she said. "She suffered greatly, and she is very sick. Since this is your area of expertise, Your Excellency, would you be so kind as to visit her for a few minutes?"

The nun, a dear friend of Maman, was very sick indeed. She was bedridden and had lost the use of her legs. Mama was very worried about her, and the doctors could do nothing. The Master was quite ready to visit the sick woman. I felt I was part of the mission, now, since I was going with him tomorrow, so I followed them to the nun's room, which was located on the third floor, with the other guest bedrooms that were always occupied by some people who needed help.

My mother knocked on the door, and the servant who was keeping an eye on the invalid opened the door. The nun looked at the Master, horrified by his bizarre appearance. "In the name of God," she said in a hoarse voice, "what is it?

Is this the Devil?" Naturally, she did not know he spoke perfect French. The Master laughed. "No, Sister. I am not the Devil." My mother smiled, and introduced them. "Sister, this is His Excellency Sung Li, a dear friend of Charles and me. Your Excellency, this is Sister Marie Ange Gabrielle..."
The nun did not respond. Generally, she was a polite, pleasant woman, and would have never insulted anyone deliberately, but her illness got the better of her and she was not quite herself. The Master apparently understood the situation perfectly.

He looked at her intently, then put his hand on the board at the end of the bed and shook it gently. The nun stared. He went on shaking it for a couple of minutes, and then said firmly, "Sister, stand up."

"He is crazy," said the nun to no one in particular.

"I am not crazy. Sister Marie Ange Gabrielle, stand up!" commanded the Master.

To my utter amazement, Sister Marie Ange obeyed. She stood up, and after swaying on her feet for a second, started walking about the room, the first time in months. Then she sat on the bed and stared again at the Master. "What happened?" she said weakly. "I walked, didn't I? But I can't walk. I am sick."

"There is really no need for you to be sick, Sister," said the Master casually. "So you cured yourself. You will have no trouble from now on."

Indeed, Sister Marie Ange got well, went back to the convent, and lived a long and healthy life. But she never quite understood what happened. Well, most people do not know how to deal with miracles, sometimes not even nuns.

*** *** ***

Leaving Paris

The next Day the Master came. I kissed Maman and Sylvie and went with him to the street, where his chauffeur waited in a large, black Citroen. The night before I lay awake, afraid that I will have a fit of crying during the parting from my family, but somehow I did not feel stressed or upset when the time came. I wonder if the Master did something to my mind, calming it down – but I never asked and I will never know.

I remember we went first to Italy, and then to Morocco, and from there boarded a ship to India. I have no clear memory of any country before arriving in Benares.

I imagine the trip was uneventful, and I probably stuck pretty close to the Master, because I was still afraid of just about anything around me. Having a strong memory for conversation, though, which later in life was extremely helpful to me, I remember a little of what the Master told me about Benares. I was naturally quite curious about it.

"It is a city of many names," said the Master. "I particularly like the name Kashi, which means 'City of Light.' The name Benares is really rather new, given to the city by the British. It's not my city of birth, of course, but I am fond of it."

"Is it beautiful, like Paris?" I asked.

"Very different from Paris, and much older," said the Master. "It is two thousand and five hundred years old. Is it beautiful? Well, some parts are, some are not. Indeed, some segments of the city are very poor, very miserable. But other parts are magnificent, and the whole city is interesting. We will have so much to show you, Germain."

Arriving to Benares

I could not wait to see it but at first sight I did not like it, because as we approached the city by train, and had to walk a certain distance to the house, we passed through incredibly poor areas. The streets were literally lined with people, lying on the ground, wrapped in their sand-colored clothes, the same color as the ground. I could not understand why people had to be so poor as to sleep in the street.

You could not even tell if the people were dead or alive, and the scene frightened me very much. But soon I realized that no one paid particular attention to the situation.

Women wearing colorful silk saris, that were as striking as the feathers of tropical birds, wove their way among the bodies on the ground. Westerners, mostly British, did the same, looking very military and imposing.

My first encounter with the paranormal
And the snake obeys the Master

Suddenly I saw a horrible thing. A large snake crawled among the people on the ground, slithering here and there. No one moved, allowing the snake to pass. I froze with terror; I never saw a snake, except in the zoo. The Master put his hand on my shoulder. "The snake will not hurt you, Germain," he said. He raised his hand, and twisted it around, making a strange sound. This was bizarre.

How could one hand make a sound? Apparently, the snake heard it. It rose vertically to the air, went down again, turned, and left the scene. Many of the poor children came to thank the Master. I realized he was well known around this area.

"You see, Germain, the snake simply went about his own business. It is not right to assume that he meant to harm anyone and be afraid of him, he had no such intention," the Master said cheerfully. All his lessons were like that. He never said, "Watch, I am going to do something wonderful now, pay attention." No, he did not want us to pay attention to himself, only to what we could see and learn. And indeed his students, knowing that, never interrupted him but always paid attention, since just being around him was a constant learning experience – and a very pleasant one at that.

*** *** ***

In the house of the Master

We finally reached his house, a very large place with a garden full of trees and flowers, incredibly lush growth which was completely different from the orderly gardens of Paris. The Master introduced me to the family and took me to my room, an extremely pleasant one, overlooking the garden. I changed into native clothes made of comfortable cotton, and joined the family to an excellent dinner.

For a few weeks, the Master had to go to his schools in Malaysia and Okinawa, and I stayed with his wife, a wonderful, kind lady whom I will always love.

The house contained a large extended family, including his wife's sister, her own son and his wife, and their five children, two of which were very close to my age and became very friendly toward me right away. In addition, the school, next door, housed more children.

Every so often a group of children from other locations, even other countries, came to stay for a while. The current group was from Tibet, all wearing saffron robes and chattering like birds.

*** *** ***

Temples, flowers, and lit oil lamps, to float on the river...

So the nephew's children took me all over Benares. At first I was terribly afraid, and didn't really want to go anywhere, but I was not going to let them know, I was too proud, so I forced myself to go and let them show me everything. The more I saw, the more interesting the place became to me.

One of the first things they showed me were the ghats. Benares is situated on the River Ganges, and everywhere there were series of paved steps that lead into the water. Most of them had tiny temples built into their sides, and people who came to bathe in the Ganges would descend the ghats and then visit the temple while still standing in the water. They often prepared flowers, and lit oil lamps, to float on the river, as part of a ritual for washing away all sins and troubles from one's life. The children told me that the ghats and their temples were extremely ancient and built by royalty.

One morning, an older cousin took us for an adventure. We took a boat, and travelled down the Ganges at the very early dawn. We covered quite a distance, which allowed us to see some of the larger temples.

One of them, a temple dedicated to the Lord Shiva, was so covered with gold plating that its towers simply glowed in the morning sun. I was surprised as to how many people were already up and about, dozens of bathers waving at us in the best natured way as we passed along, always hearing the lovely sound of the temple bells.

I was also thrilled to see the Benares Hindu University, not because I had any scholarly aspirations, at my tender age, but because my friends told me that the Master taught there. It was a huge, impressive university, established in 1916, possibly the largest in Asia at the time. I hoped the Master would take me for a tour when he came back, which he did.

No matter what day of the week, one festival or another was celebrated in Benares. I did not pay much attention to the subject of the festivals, most of which were complicated and beyond my understanding, but I enjoyed the spectacles and was seriously interested in the sweets that were sold there.

We never failed to get some – interesting tastes, very different from the sweets of Paris. My favorites were mik-based sweets, flavoured with rose water and saffron. I was also quite partial to dried fruit layered with cream and wrapped in betel leaves.

*** *** ***

The magical room of colored papers

Eventually, after a few weeks, the Master came back. We were very happy to see him, of course. I was particularly interested in his return, not just because I liked him, but because I saw a mystery or two at the house that I wanted to question him about. First, I discovered a room which had a closed door. I knew I should not enter, but I did anyway, and to my surprise I saw a very untidy mess of papers, all sizes and colors. I was not comfortable asking anyone else, so when the Master came back, I confessed my spying and asked him what the papers were for. The Master smiled, and said, "Go choose whatever paper you like, any color, any size. Bring two or three pieces."

I picked a few nice pieces and returned. The Master asked, "what would you like to see? A bird, maybe? Shall we have a bird visit us?"

"But birds don't go in houses," I said. "Only if they are lost. I don't want any bird to be frightened and lost."

"Not everything is as it seems," said the Master. "Some birds are not lost, nor are they afraid. They just visit." He quickly made a few folds in one of the papers, a white one, and to my amazement, a neat sculpture of a pigeon was sitting in his hands. I laughed, delighted with the trick.

"And I think a bird likes flowers, doesn't it?" asked the Master.

"Yes, they do," I said with conviction.

The Master made a few folds in another piece of paper, a red one. A rose magically appeared in his hand. I was thrilled, and touched the paper carefully. It was all so lifelike. As I touched the bird, it flew out of the Master's hand. I recoiled, slightly shocked.

"Nothing to be afraid of," said the Master. "Come along." He took me to the garden, and the bird flew after us in a rather business-like manner, as if knowing exactly what it meant to do. In plain day light, the Master gently put the rose on a rose bush. The paper rose immediately turned into a real flower, and the paper bird, now a real, living pigeon, settled on the bush and made distinctive pigeon sounds.

"These two are not lost at all," said the Master. "I think they are very happy." I had to agree. Somehow, the incident, despite its magical and unusual tone, did not frighten me at all. I loved it.

*** *** ***

And then came the miracle of the tree

And then came the miracle of the tree. In another room in this large house, around the exit to the garden, and with its door wide open, resided many empty flower pots, with just dirt in them but no plants. Passing by them one day, I asked the master, "Why do you keep all these empty pots?"

"They are made for giant trees that like to live inside the house," said the Master.

"But how can you fit a giant tree inside? They are bigger than the ceiling," I said skeptically.

"Well, I really was remiss in not having a few in the house already," said the Master. "They are very important and bring happiness and luck. Please choose one pot, and let me show you how the biggest tree in the world will fit into it."

"But it can't come into the house by itself," I said.

"No, they don't walk, but we can go out and look for it," said the Master. We went out and stood before a giant pepper tree, covered by feathery leaves and red tiny dots of the pepper spice. A living, thriving, beautiful tree that must have been in the garden for many years.

"This tree?" I asked. "But it is not cut, it is growing! Don't cut it, it may be hurt!"

"Of course I won't hurt the tree. But we needed to choose a special kind of tree, right? That is why we are looking at it. Now come back inside and see what happened," said the Master. We returned into the house, and in the formerly empty pot stood a tiny tree, the exact replica of the giant tree outside, complete with the small red dots of the pepper spice. I stared, speechless. How did it go into the pot?

"This tree is older than the one outside," said the Master. "As a matter of fact, the giant tree is the baby of this little one, and grew from one of its seeds. Where I am going to take you some day, when we start our serious studies, we don't measure people by their size. We measure them by this," he said, tapping the top of his nose to the top of his forehead, "and by that," tapping a small area around his heart. "Size means nothing."

*** *** ***

Miracles and tricks of illusion in the streets of India

Gradually, slowly, something was happening to me. I calmed down. Under the peaceful influence of the teacher, the magical occurrences, the friendships I developed, and the newness of the culture, I began to feel more and more confident. Of course, I was not aware of it, until an interesting incident brought it to my six-year-old attention.

That morning we encountered, at one of the festivals, an event that was new even to the nephews. A man, assisted by a child of about my age, were arranging some baskets on the ground, and a large crowd, visibly excited, was waiting for the spectacle to begin. Being small and agile, we slithered among the people and found a good spot. The man sat on the ground, and removed a cloth from one of his baskets.

He took a musical instrument, a sort of flute, and began to play. To my utter amazement, a thick rope rose vertically from the basket and started inching its way up. This went on for some time, until the rope reached an enormous height. It just stood there, entirely rigid, defying the laws of gravity.

The man stopped his playing, and the child went to the rope and started climbing it. Up and up he went, finally disappearing from view. The audience whispered to each other, mesmerized. The man started calling the child to come down. I could understand a little bit of the language by then, even though the nephews spoke English to me, and that was a simple enough conversation and I understood it perfectly. From the far height of the rope, we heard the child refusing to come down.

From one of the baskets, the man took out a huge butcher knife, put it between its teeth, and started climbing the rope. Soon he also disappeared from sight.

The audience, apparently knowing what to expect, seemed agitated to a great degree. Suddenly, a scream was heard, and horrible, bloody body parts fell to the ground. Many of the witness screamed, and one woman fainted, but no one paid any attention to her, their eyes glued to the rope. Soon the man came down, picked up the revolting body parts, and threw them into another basket.

He wiped his knife on a piece of cloth, and covered the basket that contained the body parts with it. The audience was so silent you could hear a pin drop. Then the man picked up the blood-stained cloth and flourished it in the air. The boy, safe and sound, jumped out of the basket, smiling broadly. He then proceeded to go around the audience with a bowl, receiving coins.

"It's fake," I said to the nephews, quite decidedly.

"How do you know?" said one of them. "It seemed real enough to me. Horrible, really."

"I just feel it," I said. "I was not scared. It's terribly disgusting, but it's a fake. I know it in my head."

I suddenly felt a hand on my shoulder. I turned, and saw the Master. He was smiling broadly at me. I could see he was extremely pleased, but I did not know why.

"Did you see it, Master?" I asked. "It's fake, isn't it?"

"Yes, it is, Germain. If you are interested, I will explain to you later how he did it. A mere slight of hand. But what I find interesting, Germain, is that you were not scared by the sight of it, which was really rather terrible. Have you felt any fear?"

"None at all, Master."

"When was the last time you felt afraid, Germain?"

"I am not sure, Master. It's been awhile, I think..."

"Do you remember how scared you were when you came here?"

"Yes," I said, rather intrigued. "Everything, even that silly snake, scared me. I don't know why... but I am not scared anymore."

"And when we go back to Paris, will you be afraid of the bombs, and of the nasty people?"

"I don't think so," I said, considering. "Maman said the war is over, and I am not sure I remember the bombs too well now. But even if it were not over and there were bombs... I just don't feel this sort of pain, this thing in me, I can't describe it... how did you do it, Master? How did you make me not scared?"

"I did not do anything, Germain. Not a thing. I just showed you a few beautiful little things about life, like the bird, and the tree. You have conquered your own fears, and I am proud of you. But not everyone can do it on his or her own. So my own teachers developed a system of how to cure ourselves of fear."

"Can I learn it? Maybe someday I will be scared again?"

"Certainly you can learn it, though you will never need it for yourself again. In addition, you will probably find that much of it you have done to conquer your own fears. However, some day you will teach others how to use it. Actually, it's fun."

I knew it would be, because everything the Master taught was great fun. And even though he was right, and that time in Benares cured me from trauma and fear forever, I am happy I can teach others how to acquire the skill. The interested reader can find it in Lesson One in the book, and I hope it brings relief to anyone who needs it.

*** *** ***

Return to Paris

Soon after these events we had to get ready to return to Paris. I was torn between the joy of reuniting with my family, and the sorrow of leaving my new friends and wonderful vacation. But we all knew that we will soon meet again.

For a while, the Master and I corresponded. He wrote to me about once a month, on beautiful rice paper, soft and elegant. Each page had an ornate frame around it, delicately depicted in lovely colors, and sometimes there was a picture faintly embedded in the paper itself, perhaps of cherry blossoms, or bamboo, or birds.

Sometimes, when he wanted to emphasize a concept, he drew a picture. I liked these letters so much I asked my mother to have them framed for me, and I hung them on the wall. I did my very best to write back as neatly and as carefully as I could, because I wanted the Master to see how much I was improving in school. Emulating his style, I also acquired the habit of adding little drawings to letters – which I have continued even when I grew up.

And then, to my great joy, the Master came back to Paris. He had more treaties to take care of, but that did not interest me greatly. What I was really happy about was that he asked Maman if I could go with him on another vacation, this time in Hong Kong. It was August, so we had about a month for our trip, and I could come back in September for the beginning of the school year.

"Is it safe, Your Excellency?" asked Maman anxiously. "I know Hong Kong is now in the hands of the British again, but still..."

"It will be entirely safe, Madame Lumière," said the Master. "Life returned to normal, the people have recovered so quickly from the War. Hong Kong is very pleasant these days."

"I was thinking more of the tropical storms," said Maman.

"I don't expect a very dangerous season," said the Master. "And anyway, we will always be in protected areas."

"Very well," said Mama, "but, Your Excellency, it may frighten Germain... would you be very much afraid of a tropical storm, Germain?"

I considered. "I don't think so," I said. "I was not afraid all this year, not of anything."

"It can be a good test," said the Master, looking at Mama with his pleasant smile.

"Yes," said Maman. "You are right, your Excellency. I will let him try his wings again."

*** *** ***

On the road again...on my way to Kowloon

Naturally, we stopped first in Benares for a happy reunion with the Master's family. We could only stay one or two days before taking our plane directly to Hong Kong, but it was a joy to see everyone.

We went to the Kowloon Peninsula, the most urban part of Hong Kong, and I enjoyed walking in the noisy streets and seeing the perfect blend of East and West. The streets were very busy, with bicycles, cars, rickshaws, and pedestrians all shoving and honking and screaming and bowing to each other as the need arose.

Everywhere were big buildings, like in Paris, but with some Asian flair that cannot be defined, standing next to Buddhist temples. You heard people talking in many languages, since Hong Kong was an important port, but the official languages, as I found out from the Master, were Cantonese and English, so the myriad of signs in the streets were written in these languages.
Enticing scents of food, spices, perfume, flowers, and incense floated everywhere.

On some street corners you saw Chinese shops where you could visit a doctor who also sold traditional herbal medicines. You could stop for a bowl of shark fin soup, real or fake as the case may be, and shops that sold pretty little statuettes from all over Asia. But you also came across a very English-style pub here and there, a church, or a movie theater showing the latest from Hollywood. It was oppressively hot and humid, but I did not care about such things, and the Master, able to handle any climatic condition, did not seem to mind either.

*** *** ***

The mirage island

The Master had some business to conduct in Hong Kong, but after concluding them, he told me we were about to go to a place which was unique, and where no French person had even gone.
We boarded a bark, a very simple affair that was clearly not designed for tourists, and in the afternoon of that day sailed off. After an uneventful trip, something appeared in the distance. "Is this the land we are going to?" I asked.

"Yes, we are almost there," said the Master. "It is a very small island, linked to a chain of other small islands by tiny harbors. If you look very carefully now into the distance, you will see an extremely unusual phenomenon. Concentrate on where these jagged, vertical cliffs should ease into the sea."

I looked, to my amazement, I saw the earth between the cliffs and the sea moving, shaking, as if in an earthquake. I did not notice any port, just this quivering land.

I did not know what to make of it, and continued staring at it with disbelief, until we got much closer and I realized that what I thought was land really was a fleet of small boats, each attached to the other, all the way to the tiny beach which stood at a slight incline toward the cliffs, completing the mirage-like phenomenon. It seemed like a platform made of boats. All were gently rocking on the water, their movement creating the illusion of an undulating land.

"So how do we get to the shore?" I asked, fascinated by this magical place. "There is no port and the small boats are blocking the way to the beach!"

"They are not blocking it," said the Master. "They are our bridge. We are going to walk on them, jumping from one boat to the other. Come, I will pick you up and we will fly!"

Indeed, if felt like flying. With one of the sailors bringing our small luggage after us, the Master leapt from boat to boat like a huge bird, his robes and his beard streaming in the wind, finally landing on the wet, golden sand of the beach. I laughed as we stopped and he put me down. "Master, it is as if you have sprouted wings!" I said delightedly. "Can we do it again?" The Master laughed too.

"We will, on our way back. There is no other way to catch a boat on this island. Believe me, Germain, you are the only Parisian who has ever done that," he said. "They are almost entirely isolated here. The grown-ups may have gone out of their island occasionally, but the children here never saw a white boy before!"

He was referring to a crowd of children of all ages that had already gathered around us. They looked at me as if I were an alien creature. Of course, wearing a small cap, dressed like a Parisian child, and having white skin and green eyes, I must have struck them as utterly bizarre.

Some of them moved closer, and gently touched me, as if wondering if I were real. They were very quiet and polite, though, and as I smiled at them, they smiled back and bowed. I, in turn, found the inhabitants a little strange as well, since the men, women, and children wore clothes that to me seemed very similar and I could not tell who was who. They all looked so much the same.

After a while we walked toward the village, followed by children at a small distance. The island, being a very poor place, had no electricity. Instead, they used lanterns.

As dusk fell, one lantern after another was lit, white, orange, green, red, and gold, the little lights were like shooting stars coming from all directions toward me. Being used to the lamps that were lit all at once in Paris, this gradual, organic illumination of the soft darkness was magical to me.

*** *** ***

The small temple on the top of a hill
Talking to the Buddha

Our destination was a small temple on the top of a hill. Before entering, the Master showed me the Wheels of Fortune. The colorful wheels were attached to sticks. As you passed by them, you were supposed to turn them, one by one, for luck. They began to spin, changing colors and making a soothing noise.

We entered a big room with high ceiling, the lovely scent of amber permeated the air. Candles were lit in many little niches, and everywhere stood little statues of men and women, made of green, pink, and lilac stone or metal. The whole place was incredibly colorful, full of tints I have never seen before except in crayon boxes.

Such colors were not used in Paris, or in Benares, but here they leapt out of the crayon box and into the real world. Huge masses of flowers filled the place, and the richness and liveliness of this room were like the birth of a fresh, new universe.

"You can bow before the Buddha and ask for a favor," said the Master. "It's the custom."

"Very well," I said, and bowed to a large statue. "I want to grow up, be a big president in France, and not let any Germans come to France ever again."

"So," said the Master, smiling kindly at me. "You want to be a president, an important person?"

"Yes," I said, "I want to be important. Of course I do. Doesn't everyone?"

"No, no," said the Master. "This is not a good thing to ask from the Buddha. Remember, Buddha was born a prince, but he did not want to be an important person. Instead, he wanted to do good. So he would not like such a request. The best thing to do is to start by asking favors for others."

"But this would be stupid, as I would be getting nothing for myself," I said, a bit bewildered.

"Don't you see? If you ask favors for others first, then they will ask favors for you first!"

That made sense. It may be that a large numbers of people will ask favors for me, if I start like that. I listened carefully as he continued. "This way you get rid of egoism, you don't thing about yourself first, you give first and receive later. I will show you.

Ask the Buddha that the children of this fishing village will be happy and healthy, and never go hungry." I did as he asked, and he put his hand on my shoulder and took me out of the temple.

Outside, the children, who apparently followed us all the way to the temple, were chatting or playing with some homemade toys. One of the toys was made from a stick, which held a drum made from thick paper. A small cord, ending with a sturdy knot, was attached to two sides of the drum. The child would twirl the stick, and the knots in the cords would hit the drum and make a very beautiful sound.

One of the children came forward, and unexpectedly offered me his own drum as a gift. I was delighted, and thanked him as best I could.

The children laughed and ran away, singing and chatting like many little birds. "You see?" said the Master. "You received without asking."

It was true, I thought. I asked a favor for the children, and the children gave me a toy. I cannot begin to tell how good this exchange felt. So different from the Parisian way. So lovely and kind. Later I found out that this was really a way of life in this village.

*** *** ***

The villagers, poor as they were and experiencing constant struggle for survival, were unusually loving, and happy

The villagers, poor as they were and experiencing constant struggle for survival, were unusually loving, and happy. For example, they had to live in close proximity to each other, in truly crowed conditions. Instead of being unhappy about it, they took advantage of the closeness by always knowing who needed help. Often, the cooking was communal. If someone needed food, he or she knew that soon enough a neighbor or two would know about it and share their own meagre supplies.

"We are going to stay in this temple as our lodging," said the Master.

"What? We will sleep here, with all these statues? Where?"

"No, do you see the beautiful curtain by the far wall? The one made of silk and lace? The woman who is in charge of this temple lives behind this curtain. She has rooms to let, and I have arranged in advance for us to sleep there."

We walked behind the curtains, and entered a corridor that led to a few rooms. Each was lined with wood and clay, and had a low ceiling, unlike the very high ceiling of the temple. They were very clean and neat, and after the woman gave us an excellent dinner consisting a long strips of fish that were cooked in flour, rice, and bowls of dried fruit. As soon as I finished eating, I sank into one of the very comfortable beds, exhausted by my adventurous day, and slept immediately.

I woke up very late; wishing to find the bathroom, I wandered here and there and could not find it. I was annoyed, but luckily as I walked out the Master was there. "Are you looking for the bathroom?" he said.

"Yes, I went all over the house, there is no bathroom!" I said.

"They won't have a bathroom in the house," said the Master. "You fell asleep so quickly I had no time to tell you. All the bathrooms are in the yard, since they consider it unclean to have one in the house." I shook my head, wondering about the difference of civilizations from each other, and went in the direction the Master pointed at.

*** *** ***

Feeding the fish with food from heaven

"We will go about the island and show you how people live here," said the Master. We went down to the small beach. Some boats were on the beach, turned upside down, and the men were repairing them.

Unfortunately, the Master explained, they are so poor, and have so little wood, that sometimes they had to simply patch a boat rather than fix it properly, and endangered themselves when they went to sea. But they had no choice in the matter. Others were fixing their fishing equipment. I saw that when they fished near the beach, they used nets, and when they fished farther in the water, they used boxes made of rattan or bamboo. "Would you like to try to fish?" said the Master.

"I have never fished before," I said. "What would you use, a net or a box?"

"Neither," said the Master. Went to the edge of the sea. Soft little waves touched the rocks that lined the beach like a natural pier.

"You can take your shoes off so they won't get wet, but don't put your feet or play in the water for a little while," said the Master. "We don't want to scare the fish." He sat on a rock, rolled up his sleeves, and put his finger in the water. He held nothing in his hands. I watched, fascinated, as the fish started to come to his finger, stuck their little heads up and opened their mouths.

From somewhere, I can't imagine where, the Master produced quite a lot of crumbs, and fed the fish. That was the Master's idea of fishing; he never killed an animal in his life.

*** *** ***

The secret is how to use your hands properly

When he was done, and the fish left, I grabbed his hand to see if there was any mysterious object in it, something with which he called the fish to his fingers.
But there was nothing in his hands, nothing in his sleeves.
The Master laughed. "No, I have nothing in my hands, Germain. It is simply knowing how to use your hands properly. Do you know how to join your hands?"

"Yes," I said, and grasped my hands together.

"There is a better way," said the Master. He leaned his closed fist inside his other hand, which was open. "You see, Germain, the fist, which you make with the right hand, is for strength and power. The left hand, left open, is the shield. By holding the fist and the shield together, you protect others from your own aggression."

I tried it, and he approved of the way I placed the hands. "Now," he said, "I am going to teach you how to use the hands, followed by your body, to do what we call 'ballet with nature.' I would like you to practice it every day. Would you do that? And later, when we meet again, I will show you the next step."

"Sure," I said. "How do you ballet with nature?"

"First, you move your left arm in circular motion. Let your body flow with it, your whole body; sway and turn with it. When you get tired, move the arm in the same way, only in the opposite direction."

I tried, and it felt very nice both ways.

"Now, take the right hand, make a fist, and repeat the same dancing motion, first in this direction, then the other." I tried, and again, found it easy. Then I combined the two motions, under his instructions, and he approved. I promised I would do it every day, which I faithfully did. I honestly thought I was dancing... only later I found out, when it was necessary and urgent, that the Master, the gentle, loving Master, really taught me the first rules of deadly self defense, and it would be rather handy later in life.

*** *** ***

The Master: "You will find that animals understand more than people do, if you love them."

In the meantime, we returned to the temple. There were a few birds on the shore, not many. To my surprise, they took a look at the Master, and a couple of them flew straight at him and sat on his shoulder. The Master produced more crumbs from thin air, and the birds had their lunch. How did they know he could feed them, I wondered. "They just know," said the Master. "You will find that animals understand more than people do, if you love them."

I have learned so much and enjoyed my stay in the island, but unfortunately we could not stay as long as we wished. The news came that the weather was changing and tropical storms were forming not too far away.

We retraced our steps, flying happily over the little boats again. I enjoyed it so much that I did not notice that the sky was turning a rather ugly, greenish gray. We boarded our boat, and headed to the port. The wind started to rise, the sky was darker and darker. I looked at the Master and asked, "Is that the big storm?"

"Yes," said the Master, looking intently at me.

"I see the waves are becoming rather high," I commented in a matter of fact way.

"Indeed," said the Master. "It may develop into a rather nasty tropical storm."

"How interesting," I said. "I will tell everyone at school that I sailed during a typhoon. They will be so jealous." The Master started laughing.

"What is so funny?" I asked, surprised. "The children in Paris don't have a lot of adventures, like me."

"Are you not afraid, Germain?"

"No, not at all," I said. "I am sure the sailors have done it a thousand times before..."

"Well," said the Master, "we have accomplished a great deal, Germain. You have lost the last trace of your childhood fears. I am proud of you."

Ballet with nature

But this was not entirely the end. There was one more proof I needed for myself, and that happened a few years later, when I was already in high school. A group of bullies, who did not like my political views, or anything else about me for that matter, surrounded me in the schoolyard.

For a moment a pang of fear struck through me, but it did not last. I was suddenly cold and calm, and for reasons I did not at the time understand, I started doing the "ballet with nature" the Master taught me on the island by the sea.

I waved my arms quite correctly, and then my right hand, made into a fist, shot into the face of the leader. In an instant, he was on the floor, his nose bleeding profusely.

The rest lunged at me, but my arms, entirely on their own, waved in the air and one by one they joined their leader, except for those who ran away. That was the last proof I ever needed. I have never, since this incident, experience fear.

We have reached the shore with no trouble. The storm veered and turned back into the sea, and we stayed for a couple of days and then went back to Paris. I was sad on the plane, since I had a feeling it was going to be for a while before the Master will come back for me.

But the Master said, "Don't be unhappy, Germain. We will always see each other again, and no matter what, I will always know where you are. Come, let me show you how to make paper flowers, like the rose I made for you in Benares. You can give it to your mother when you get home." This cheered me up. "I will also make one for Sylvie, and one for Aunt Pauline," I said.

And so I learned how to make my first origami, hoping that some day I will learn how to turn them into real flowers, just like the Master.

*** *** ***

Chapter 55
The Unthinkable Horrors

- Victoria (Sinhar Ambar Anati) in her own words...125
- How I was taught the truth about alien abductions...125
- The unthinkable horrors...125
- What do the Grays want of us on earth? ...127
- The Anunnaki monitor...127
- The room was full of operation tables, and human beings, each attached to the table and unable to move...128
- The needles reached every part of the human bodies, faces, eyes, ears, genitals, stomach...129
- The Grays can enter the human mind quite easily...129
- Most of the Grays have no substance, and therefore they are invisible...131
- "This is a baby's head! A floating baby's head! What in the world it is attached to?" ...132
- Many of the abductees who are thus psychologically influenced fall into a pathological attachment to the hybrid child...134
- Some look reptilian, some insect-like. Some eat human beings and other sentient species...134
- At the Grays' hospital ward: Inside floated a large number of severed arms and legs, all human...136
- Each tank filled with body parts, some even with full bodies...136

- The smell of formaldehyde was so intense that I almost fainted...137

*** *** ***

Chapter 55

The Unthinkable Horrors of the Grays

Excerpts from the book "Anunnaki Ultimatum"
Chapter Six of the Book
The Unthinkable Horrors
Co-authored by Maximillien de Lafayette and Ilil Arbel

Anunnaki-hybrid Victoria (Sinhar Ambar Anati) in her own words...

> How I was taught the truth about alien abductions, saw a monstrous experiment, viewed a non-physical demonic dimension, and visited the horrific laboratory of the Grays.

We were going to our wedding trip in Paris in a few days. Marduchk had some business with the Council, regarding a new assignment which he planned to work on after our return, so I had a few days to settle down and prepare for the trip.

But there was nothing to prepare, really, since everything was ready, packed, and arranged in the spaceship in a few minutes. Therefore, I decided to take Sinhar Inannaschamra at her word and make a little visit to the Academy.

I asked her if it were convenient, while she was having dinner with us one evening, and she was delighted to oblige.

The next morning I went to the Academy. It was a huge, sprawling complex of buildings, all of them connected in the traditional Anunnaki architecture, and when I first saw it, during my previous visit, I remember thinking how amusing it was that even on another planet, places of higher education had the same look as our own universities.

The gray stones, the ivy-like climbers, the winding walks among ancient trees. You simply could not mistake it for anything else but a place of learning. Sinhar Inannaschamra took me to her own office, which was a very pleasant library with thousands of conical books and desks cluttered with more cones and ancient manuscripts.

The light that filtered through the climbers by the windows tinted the air with green, and soft armchairs made you wish you could settle for a good reading session that would last for hours.

Sinhar Inannaschamra was pleased with my admiration of her place. "Yes, I love working here," she said. "I cannot imagine myself working anywhere else. I have been here for ages, ever since I have decided on my mission, which in my case was teaching future generations about the history of the Anunnaki and its relationship to the universe.

And the more I teach, the more I learn, and the more I realize how infinite is the field... which brings me to a question I really wanted to ask you but did not think, yesterday, that it was a topic for a dinner conversation. Did you devote some thoughts to your own future mission?"

"Not really," I said, a little embarrassed. "Miriam said I should not rush into it. She suggested I concentrate on the four parts of the ceremony, then the trip, then settle down for awhile, and only then start thinking about it."

"Oh, of course, there is no rush whatsoever," said Sinhar Inannaschamra. "I was thinking about it merely because if you have something in mind, I would start your studies by introducing you to it, rather than go randomly at various subjects."

"Well," I said, considering. "I know one thing. I am a beginner, a stranger, and rather young. I don't think I would do very well with species I am not familiar with, and I don't think I can feel comfortable, at this stage, with shape-changing. In addition, I have a very strong interest in the fate of the hybrid children on earth, considering that this was what I was asked to do when I had just started my relationship with the Anunnaki."

"Do you think, then, that you may be willing to work with the hybrid issue? Some of it is very nice, like your situation, because the Anunnaki treat the people they contact with utmost respect. But other species are not exactly like that. Some of it is even gruesome."

"I think I can only make such a decision if I am to know what really happens with the hybrids," I said. "Even if it is gruesome, I am aware that not everything I will encounter in my life here is wine and roses, even though it sometimes does seem like a paradise."

"You display an excellent attitude," said Sinhar Inannaschamra. "I am sure that if I show and tell you a few things, you will have a better base on which to form an idea and eventually to make a decision. The first thing we must talk about is the situation with the Grays."

"The small extraterrestrials with the bug eyes, right?" I said.

"Yes. They do not contact people like we do, on your planet. They abduct them. You are probably familiar with many stories that come from the people that they have abducted, but much of what these people say is inaccurate, and based upon mind control that the Grays exercise on them. I am going to tell you a bit, and then, if you feel up to it, we will take a short trip on my spaceship and visit one of their labs."

"Would they let you in? Aren't they dangerous?"

"Dangerous? Very. But not to an Anunnaki. We are much stronger and they are afraid of us. If I come to their place and demand to see a lab, I will see a lab. In addition, I want to show you a few things on a monitor. Some will be extremely unpleasant, but it cannot be avoided if you want to learn something."

*** *** ***

What do the Grays want of us on earth?

"What do they want of us on earth?"

"There are a few things that they want. First, they want eggs from human women and use them to create hybrids. Let's take a look at this monitor, and I'll show you how they do that. But Victoria, steel yourself. This is pretty horrible, even though I have seen even worse. You will also be able to hear, it is like a television."

The Anunnaki Monitor

The monitor blinked and buzzed, and a small white dot appeared on the screen. It enlarged itself, moved back and forth, and settled into a window-like view of a huge room, but the view was still rather fuzzy. I heard horrendous screams and froze in my seat; these were sounds I have never heard before.

After a few minutes the view cleared and I saw what seemed to be a hospital room, but it was elliptical, not square. Only part of it was revealed, as it was elongated and the far edge was not visible.

The walls on the side were moving back and forth, like some kind of a balloon that was being inflated and deflated periodically, with a motion that made me dizzy; they seemed sticky, even gooey.

*** *** ***

The room was full of operation tables, and human beings, each attached to the table and unable to move

> The room was full of operation tables, of which I could see perhaps forty or fifty, on which were stretched human beings, each attached to the table and unable to move, but obviously not sedated, since they were screaming or moaning.
>
> Everyone was attached to tubes, into which blood was pouring in huge quantities.
>
> I noticed that some of the blood was turning into a filthy green color, like rotting vegetation. At the time I could not understand what that was, but later that day I found out.
>
> This blood was converted to a suitable type for some of the aliens that paid the Grays to collect it, and it was not useful in its raw condition.
>
> The creatures who operated these experiments were small and gray, and they had big bug eyes and pointy faces without any expression. I thought they looked more like insects than like a humanoid species.
>
> They wore no clothes, and their skin was shiny and moist, like that of an amphibian on earth. It visibly exuded beads of moisture which they did not bother to wipe away.

The needles reached every part of the human bodies, faces, eyes, ears, genitals, stomach

Each operating table had complicated machinery that was poised right on top of the person who was strapped to it. On some of the tables, the machinery was lowered so that needles could be extracted from them automatically, and the needles reached every part of the human bodies, faces, eyes, ears, genitals, stomach.

The people screamed as they saw the needles approaching them, some of them fainted. Many of the people were already dead, I could swear to that. Others were still alive but barely so, and some had arms and legs amputated from their bodies. It was clear that once the experiment was over, every single person there would die.

I don't know how I could continue to look, but somehow I managed. I looked at the ceiling of this slaughter house and saw meat hooks, on which arms and legs and even heads were hanging, like a butcher's warehouse. On the side of the tables were large glass tanks where some organs were placed, possibly hearts, livers, or lungs, all preserved in liquids.

The workers seemed to be doing their job dispassionately and without any feelings, moving around like ants and making buzzing sounds at each other as they conversed. They were entirely business-like and devoid of emotion. At least, their huge bug eyes did not convey any emotion to me, neither did their expressionless faces.

I watched until I could no longer tolerate it, and finally covered my eyes and cried out, "Why don't you stop it? Why don't you interfere?"

The Grays can enter the human mind quite easily

Sinhar Inannaschamra turned the monitor off. "This event is a record from decades ago, Victoria. It is not happening now as we look. And even though often we do interfere, we cannot police the entire universe or even the entire earth. They know how to hide from us. And you must understand, that often the victims cooperate with their abductors."

"Why would they?"

"Basically, through mind control. The Grays have many ways to convince the victims. The Grays can enter the human mind quite easily, and they find what the abductees are feeling and thinking about various subjects. Then, they can either threaten them by various means, or persuade them by a promise of reward."

"Reward? What can they possibly offer?"

"Well, you see, they show the victims images through a monitor, just like this one. They tell them that they can send them through a gate, which is controlled by the monitor, to any number of universes, both physical and non-physical.

That is where the rewards come in. For example, if the abductees had originally reacted well to images of Mary or Jesus, the Grays can promise them the joy of the non-physical dimensions. They show them images of a place where Mary and Jesus reside, where all the saints or favorite prophets live, and even the abode of God. They promise the abductees that if they cooperate, they could live in this non-physical universe in perpetual happiness with their deities. Many fall for that."

"And if they resist?"

"Then they show them the non-physical alternative, which is Hell. Would you like to see some of it?"

"You can show me Hell?" I asked, amazed.

"No, there is no such thing as Hell... it's a myth that religions often exploited. But I can show you what the Grays show the abductees, pretending it is Hell; they are quite devious, you know. You see, some creatures live in different dimensions, where our laws do not apply. Sometimes, they escape to other dimensions.

These beings have no substance in their new dimensions, and they need some kind of bodies to function. At the same time, the Grays can tap into numerous universes, because they can control their own molecules to make them move and navigate through any dimension. Well, a cosmic trade had been developed. The Grays supply the substance taken from human abductees, and from the blood of cattle. You must have heard of cattle mutilation, where carcasses of cows are found in the fields, entirely drained of blood? The Grays do it for their customers."

"How do these creatures pay the Grays?"

"By various services. Once they get their substance, they are incredibly powerful in a physical sense. The old tales of genies who can lift buildings and fly with them through the air were based on these demons; the Grays often have a use for such services. But let me show you a few of these creatures. Of course, you can only see them when they have already acquired some substance from the Grays."

Most of the Grays have no substance, and therefore they are invisible

The monitor hummed again as Sinhar Inannaschamra turned it on. The white dot expanded into its window, which now, for some reason, was larger and took over the entire screen. All I could see was white fog with swirls floating through it. Sometimes the fog changed from white to gray, then to white again.
I started hearing moans. Not screams, nothing that suggested the kind of physical pain I saw before, but perhaps just as horrible, since they voices where those of hopelessness, despair, and emotional anguish.
Every so often I heard a sound that suggested a banshee's wail, or keen, as described in Irish folklore.
"It will take a while for someone to show up," said Sinhar Inannaschamra. "Most of them have no substance, and therefore they are invisible. Others have a shadowy substance. Then, there are the others... but you will see in a minute. Once they notice they are being watched, they will flock to the area, since they are desperate to get out. Incidentally, it was never made quite clear to me how they produce sounds without bodies, we are still trying to find out what the mechanics are, but it's not easy, because we would rather not go there in person."
"They sound horribly sad," I said.
"This is what makes it so Hell-like. In many cultures, Hell never had any fire and brimstone and tormenting devils, but rather, it was a place of acute loneliness, lack of substance, and alienation from anything that could sustain the individual from a spiritual point of view. Think of the Greek Hades, or the ancient Hebrew Sheol, before the Jews made their Hell more like the Christian one. Look, here comes the first creature. Poor thing, he is a shadow."
I saw a vaguely humanoid shape in deep gray. It seemed to have arms, which it waved in our direction. It was fully aware

of the monitor. Then another shadow, then another, all shoving each other and waving desperately at the monitor.

Then something more substantial came into view, and I jumped back as if it could reach me. It seemed to be a severed arm. Cautiously, I came back, and then saw that the arm was attached to a shadow body. I looked at Sinhar Inannaschamra, speechless, and she said, "Yes, here you see one that managed to receive an arm. It wants to complete its body, of course, so that it can get out of this dimension and serve the Grays, but the Grays keep them waiting until they want them."

More and more came, clamoring for attention. "Do they think we are Grays?" I asked.

"Yes, they do. They can't tell the difference, all they know is that they are watched, and they try to get the attention of the watchers. It's incredibly cruel, but if you feel for them, which I still do, remember that at the same time they become murderous, cruel creatures themselves as soon as they escape their dimension and join the Grays."

"This is a baby's head! A floating baby's head! What in the world it is attached to?"

Another half thing came into view. It had eyes stuck in the middle of a half-formed face, each eye different. The face seemed mutilated, somehow, until I realized it had no nose and no chin. Floating heads, arms, legs, torsos – they all jostled in front of the monitor, each more horrible than the other. And then I saw the worst thing imaginable.

"Sinhar Inannaschamra!" I screamed. "This is a baby's head! A floating baby's head! What in the world it is attached to?"

"Another shadow," said Sinhar Inannaschamra. "They don't care what age the substance comes from. Sometime the babies' heads or limbs get attached to big adult bodies."

"They use babies," I said, sobbing. "Babies..."

"Yes, this is the kind of creatures we have to tolerate," said Sinhar Inannaschamra sadly. She looked at me and realized I could not take any more of this Hell, and so turned off the monitor.

"This was something," I said, shivering and trying to recover.

"Indeed," said Sinhar Inannaschamra. "So you see, they can easily show them horrific pictures of Hell, enough to frighten them to such an extent, that they are sure to obey. Interestingly, the abductees, under such threats, often develop physical, psychosomatic effect in the form of scratches, burns, or even stigmata, on parts of their bodies.

Of course, sometimes the Grays burn them with laser beams as a form of punishment or of persuasion, and sometimes the wounds are produced simply by the radioactive rays emanating from the spaceship, like what sometimes happens in nuclear plants on earth, or during nuclear explosions. But most often it is the mind reacting to the image."

"How horrible..." I said weakly.

"It gets worse," said Sinhar Inannaschamra. "They can show them the physical universe as well. They would project, on screen, well-known events that occurred during times in which humanity was utterly cruel, or when war, famines, and plagues ravished the earth.

They might show them the Crusades, or Attila the Hun, or the Nazi concentration camps, or the famine in Ireland, or the black plague in Europe, and threaten them that they can open a gate through the monitor, and abandon them there for the rest of their lives."

"The poor things. No wonder they obey," I said.

"Yes, you see, the Anunnaki tell you the truth when they contact you. They let you know that they cannot change the past, nor can they interfere with the future. But the Grays lie. They tell the abductees that they can change events in the past, from day one, and that they could project and change life at two, five, or ten thousand years in the future of humanity."

"I wish you would just wipe them of the face of the universe," I said.

"We don't do such things. Some of us recommend it, but we just don't. Anyway, they have other systems of persuasion. Some women have a very strong reaction to the images of children. The Grays catch it, of course, and then they tell the women that they have been abducted before, years ago, and were impregnated by the Grays.

Many of the abductees who are thus psychologically influenced fall into a pathological attachment to the hybrid child

Then they show them a hybrid child and tell them that this is their own child. Many of the abductees who are thus psychologically influenced fall into a pathological attachment to the hybrid child. Then the Grays tell them that if they don't fully cooperate, they will take the child away. The woman cooperates, the experiment takes place, and then the Grays make them forget the child and place them back in their beds at home. Usually, some vague memory remains, since the Grays don't care about the well being of their victims and don't bother to check if the memory is completely cleaned out."

"But it is not really their child?"

"Sometimes it is, sometimes it isn't. You must realize that normal impregnation and the nine months of carrying the baby does not occur. The Grays take the eggs out of the woman, the way you just saw it on the monitor, put them in a tube, fertilize it by an electric or sometimes atomic way, and the hybrid grows in the tube until it is of term. No woman has ever given birth to a hybrid."

"This is beyond words," I said.

"And I thought most of the extraterrestrials would be like you."

"There are many species," said Sinhar Inannaschamra. There are those humans call 'The Nordic,' they look and behave much like humans, they are rather kindly, and other kinds are reasonable as well. But there are a lot of horrible species. The closer any species get to the demonic dimension, and particularly if they trade with them, the worse they become.

Some look reptilian, some insect-like. Some eat human beings and other sentient species

Some look reptilian, some insect-like. Some eat human beings and other sentient species, in what we see as almost cannibalistic behavior. Some make sacrifices of sentient beings to their deities. The reptilians have a specialized digestive system. They don't eat solid food, but only suck blood through pores in their fingers.

That is why some researchers on earth connect the extraterrestrials with vampirism. The Grays sell them cattle blood, since the reptilians don't particularly care where the blood comes from. But anyway, are you ready for your field trip? Let's go and visit a Grays' lab."

I thought I was ready. I thought I was tough. But what I saw on this field trip would remain with me for eons.

Sinhar Inannaschamra took me to her spaceship, and informed me that the trip would be very short. She had been to this lab before, and knew the conditions very well. Just before we landed, she pulled out a suit that was needed to protect me from any radiation.

Apparently, for this trip, she needed no protection herself, but she could not tell as yet if I could tolerate such conditions or not, due to my human existence for the last thirty years. The suit was made of lightweight, soft metallic material that was actually rather comfortable and moved easily with me. Then, I put on a helmet, which was entirely transparent and allowed me perfect vision.

We landed on a bleak field covered with some material that looked much like cement, gray and unpleasant, but with a smoother finish. Right before us was a huge building which looked like an ugly airport hangar, completely utilitarian without any ornamentation. The entire area around it was an empty prairie-like field with stunted, grayish vegetation, stretching into the horizon without any feature like a mountain or a city. The sky was gray but without clouds. We walked to a large door, tightly closed and made of metal.

Sinhar Inannaschamra put her hand on it, and it slid immediately to the side and allowed us to come in. "They know my hand print," she said to me. We entered a small hall, empty of any furniture, and from there, a door opened into a long corridor, brightly lit and painted white. On each side there were doors, also painted white, all closed, and it was entirely empty of any occupants. Sinhar Inannaschamra lead the way to one of the doors, and again placed her hand on it.

The door slid open silently, and we entered an enormous room. It was so huge that I could not see the end of it, and had gray walls and a white ceiling. Round, bright lights of large circumference were placed in the ceiling, emanating a very strong illumination.

At the Grays' hospital ward: Inside floated a large number of severed arms and legs, all human...

The impression the room gave was that of a hospital ward, or a surgical hall, but there were no beds or operation tables, only large tanks containing some objects I could not as yet see. And while the place was scrupulously clean, the smell was nightmarish. I recognized the stench of formaldehyde, mingled with some other malodorous liquids. I was surprised I could smell anything through the helmet, but Sinhar Inannaschamra explained that they deliberately made the suits and the helmets allow as much interaction with the environment as possible.

"It smells like that because this is the warehouse, where they keep all the spare parts," said Sinhar Inannaschamra. "Come, look at this tank."

Each tank filled with body parts, some even with full bodies

We approached an enormous tank, transparent in the front parts and increasingly opaque as it extended further into the room. Inside floated a large number of severed arms and legs, all human. I recoiled in horror, but quickly recovered. We went to the next tank.

It was arranged in the same manner, but inside floated severed heads and torsos. And so it went on, each tank filled with body parts, some even with full bodies. Smaller containers had interior organs, such as livers, hearts, and some others I was not sure of. In addition, there were containers of blood, some red, some green. I already knew that the green blood became like that after preparation for sale to species that needed the adjustments.

Suddenly I heard sounds of conversation, as if a group of people were approaching us from somewhere. The sounds were in a language I could not understand, and uttered in a metallic, screeching way that was almost mechanical.

To me, it sounded demonic and inhuman. A group of five Grays approached us.

After what I have seen on the monitor, I was about to escape in terror, but the group bowed to Sinhar Inannaschamra, and went on about their business. One of them approached a tank.

He looked at what seemed to be a chart, like a hospital chart but in a language I could not understand, that was positioned above the tank, and just stared at it for a short while. A line on the chart lit up, and some equipment that was build above the tank came down, entered the tank, and using a robotic hand pulled out a specimen and placed it in a tube, along with some of the liquid. Then the robotic hand came up, moved forward, and handed the tube to the Gray. The Gray took the tube, looked at the chart, and the robotic hand withdrew.

The Gray took the tube to a wall and placed the tube against it. To my amazement, the wall sort of swallowed the tube and it disappeared.

I looked at Sinhar Inannaschamra for explanation, and she said, "The walls are not solid. They look solid, of course, but really they are constructed of energy. The Grays can move things back and forth, and even pass through it themselves. Some of the walls contain drawers, where they place equipment."

The other Grays were all communicating among themselves in their demonic language, mostly ignoring us. "Let's go to the next room, where I can show you what they do with their specimens," said Sinhar Inannaschamra. We went to the wall and she put her hand on it while holding me with her other hand, and I found myself passing through the wall as if it was made of thick molasses.

The smell of formaldehyde was so intense that I almost fainted

The room we entered was designed just like the others, architecturally, but had work tables instead of tanks and containers. Hundreds of Grays stood there, each at his table, doing things to the limbs, torsos, and blood. The smell of formaldehyde was so intense that I almost fainted. "Here," said Sinhar Inannaschamra. "Let me adjust the helmet so you don't have to smell the liquid." She did something at the back of the helmet, and I felt better.

"And now, let's go to the area where they fit the spare parts on the creatures I have shown you, the ones that want to buy substance," said Sinhar Inannaschamra.

We stepped into a third room, again through the wall, and this was a much smaller room. On the walls, there were a number of monitors, just like the ones in Sinhar Inannaschamra's office, but much larger.

Before each monitor stood a Gray. We walked to one of them, and the Gray bowed to Sinhar Inannaschamra, and returned to his work.

The Gray adjusted something, and a swirling shadow attached to one leg appeared on the screen. Behind him were numerous other shadows, but the Gray managed to separate the first creature from the others with some walls of energy that looked like white fog.

The creature waved desperately at the Gray, who had before him a torso in liquid. A large robotic hand came from above the monitor, picked up the torso, and allowed the liquid to drain into its container. Then, he passed the torso through the screen, which now I realized was made of energy, like the wall, and placed it on the shadow. The shadow shivered, as if in pain, and I heard a deep moan or sigh, as the torso attached itself to the swirling gray form. I had a glimpse of a shadowy face, contorted in agony, but whether it was physical or mental pain I did not know. The shadow seemed utterly exhausted by the bizarre procedure, and floated away.

"What will happen to it now?" I asked.

"It will be back again and again, and when the Gray decide, they will give it more parts, allowing it to adjust and become more substantial."

"Are there any of them ready to leave and serve the Grays?"

"Yes, but I think this would be too dangerous to visit in person. The demonic creatures don't have the restraints the Grays have, regarding the Anunnaki, and often they are too stupid and just lash out as soon as they are brought into our dimension. Their transport, therefore, is done in a different part of the lab, under tremendous precautions. Besides, we don't want to stay much longer, since I think you had seen enough for one day..." I could not agree with her more, and we retraced our steps back to the spaceship.

"Some time I will show you how the creatures are taken out and put to service," said Sinhar Inannaschamra. "Only not in person." This sounded good to me. I was already so shaken from my day's adventures, I did not think I could take much more instruction, nor did I have a wish to meet such creatures in person. But, of course, I knew that one day I would have to do exactly that.

*** *** ***

Chapter 56
At the Anunnaki's Academy in Nibiru

- At the Anunnaki's academy…143
- In Victoria's own words…143
- Anunnaki's orientation program…143
- The lessons are entered directly into the brain's cells of the students…144
- The Anunnaki beautiful classroom…145
- The Nif-Malka-Roo'h-Dosh ritual…149
- The creation of the mental Conduit…151
- The "Double" and "Other Copy" of the mind and body of the student…151
- The Anunnaki have collective intelligence and individual intelligence…152

*** *** ***

Chapter 56
At the Anunnaki's Academy

In Victoria's own words...
Anunnaki's Orientation Program

> I am given an orientation at the Academy in preparation for discussing my mission, and I undergo the purification and the creation of the all-important mental Conduit.

It took some time, and many discussions with Marduchk, but I finally managed to put the trauma of the fight with the ancient Anunnaki behind me. It is never easy to adjust to a new culture. It is even more difficult to do so when your own spouse is part of this new and incredibly different culture. Think what it is like if you are adjusting not only to that, but to life on another planet.

I have to admit that I succumbed to a form of depression. It was not the danger that so much affected me, but I was wondering if I really understood Marduchk.

My commitment to him never wavered – I am too much of an Anunnaki to have doubts about my husband once I have made my choice – but I felt helpless and alone and terribly inadequate.

However, I received much help from Marduchk, who understood my feelings completely and was willing to support me through this hard time, and also from Miriam and Sinhar Inannaschamra, who advised me of how to manage my feelings from the female point of view. In addition, the spiritual value of the Mingling of the Lights makes a person more and more attuned to the feelings and thoughts of his or her spouse. I grew to understand Marduchk's motives and behavior better every day. And one morning I woke up to feel that life seemed good again, and I was ready for whatever was in line for me next.

The lessons are entered directly into the brain's cells of the students

"Time to go to work," I said to Marduchk over breakfast. "I think it will be good to start thinking about my mission."

"In this case, then, I think you might want to start your orientation at the Academy. You can't do a thing without it," said Marduchk.

"Just what I would love to do," I said enthusiastically. "Shall I go to Sinhar Inannaschamra and see about the test for admission into the orientation?"

"You have already passed the test," said Marduchk.

"But I only identified one ancient Anunnaki skull," I said, surprised. "You said I had to identify two."

"You had to go through an extended fight with a threatening, evil life form, traveled to other dimensions, rejected the attempts of a monster at possession, and survived his locking you at this deathtrap of the box-like room, not to mention seeing your husband destroy a Master Being and all his versions for the first time," said Marduchk in his matter of fact way. "It counts as identifying a second skull, I should think... anyway, Sinhar Inannaschamra thinks you passed, and that is all that matters. She said to me that as soon as you are ready to start, you will be most welcome to the orientation."

"How thrilling! When shall I go?"

"Tomorrow will be fine, since there are never any official dates for that, as they have on earth. I'll just tell Sinhar Inannaschamra."

"Fantastic. Do I need to bring anything with me? Supplies, copybooks, etc.? Or do I get the supplies there, on the spot?"

Marduchk laughed. "No, there is no need for supplies. The students do not have to write or copy anything. No pens or pencils required, no copybooks. We don't even have desks in the classrooms, come to think of it.

You see, the lessons are entered directly into the brain's cells of the students, and become a permanent part of their learning acquisitions."

"How in the world would they do that?"

"It's hard to explain. It will be a lot easier for you to follow it step by step and get the background information as you go along. I could tell you that the acquisitions, or depot of knowledge, will become an integral part of the intellectual, mental-scientific program of the students, but it may not mean much to you until you experience it, I suspect."

"Indeed it means nothing at all... I'll just take everything as it comes, and most likely enjoy it enormously. So, in the meantime, would you kindly manifest another cup of coffee for me before you go off to the Akashic Library? After I drink it, I must go and consult Miriam. She can help me decide what to wear for my day in school. I can't an imagine Anunnaki having to wear a school uniform."

"A uniform? An Anunnaki wearing a uniform? How funny!" said Marduchk, amused and mildly shocked by the idea.

"It might interfere with free will, won't it?" I asked, a little maliciously.

Marduchk laughed. He had begun to notice his own extreme attachment to the notion of free will, not as yet entirely shared by his wife.

The Anunnaki beautiful classroom

The next morning I went to the Academy. I have always loved this magnificent edifice, with its sprawling complex of buildings, all connected to each other in the traditional Anunnaki's architecture.

I never failed to be amused by the similarity of all places of higher education, even on another planet. You could have taken this complex, with its lovely old stones and ivy-like climbers, just as it was, and place it next to Harvard, Oxford, or La Sorbonne, and it would fit right in.

A few other students were approaching the particular building where Sinhar Inannaschamra held her classes, and we were all welcomed very cordially by one of her apprentices and sent to our various classrooms.

I entered a large, beautiful room. Later I found out it was 220 meters long by 70 meters wide,[1] dimensions that represent very elegant proportions like all Anunnaki architecture. Huge windows allowed in delicate green light, filtered through the

---

[1] Approximately 772 X 230 feet.

climbers. All the wall space between the windows was covered with shelving, on which many conical books were arranged.

The classroom was divided into three sections. The apprentice that was my host explained the section system to me. It was rather complicated, since the sections were not only constructed for the convenience of teaching, but had deep symbolic meaning. He started by asking me a question that at the time I found very strange. "Are you, by any chance, a Freemason?" he said.

"No, why?" I asked, intrigued.

"Well, you see, if you were a member of the Freemasons, much of what I have to tell you about the symbolism would have already been known to you."

"Really? How come?"

"The Freemasons, with their extremely ancient Phoenician heritage, have acquired much information from the Anunnaki, and it is still in use by them. But no matter, in no time at all you will know more than any of them.

I will fully explain the matter to you, and I might as well tell you about the Freemason and Phoenician connections," said my guide pleasantly.

"After all, you are from earth, and I understand you are extremely interested in history and archaeology." He appeared to be highly knowledgeable and extremely happy to share information, traits that seem to serve the Anunnaki well.

"Indeed I am," I said. "Thank you so much for your instructions."

"My pleasure," said the apprentice, and launched into the most informative discourse.

"Section one is named Markabhah-Ra, and its code is '3'. There are three rows, each seating one beginner, since this section is reserved for new students or recruits. In freemasonry, the code of the new recruits is also '3'. Markabhah-Ra in Anakh[2] means 'traveler.' It is exactly the same name given to a new recruit or member in a Freemason lodge. In the ancient Hiram[3] Brotherhood Society, number '3' was originally number '1.' Later on it was changed to '3' because number '1' was considered the

---

[2] The name of the Anunnaki language
[3] A Phoenician king, mentioned in the Bible

first attribute of Baal,[4] henceforth no human was worthy of receiving this sacred number.

Section two is named Kah-Doshar-Ra, and its code is '18.' There are eighteen rows, each seating one apprentice. It is reserved for the mid-level Anunnaki's students. Kah-Doshar-Ra in Anakh means Holy, or more exactly, 'Students of the holy energy or source of life.'

In Freemasonry, it is exactly the same thing. Mid-level Freemasons receive the degree '18.' With them, Kah-Doshar-Ra became Kadosh. Many Semitic and ancient Near East/Middle East languages have a word much like it, including the Hebrew and Aramaic Kadosh, the Coptic Kakous or Kouddoup, the Syriac Kouddous and the Arabic Kouddoup or Moukadass.

Section three is named Wardah-Doh-Ra, and its code is '33.' There are thirty-three rows, reserved for the Masters, each seating one Master. Wardah-Doh-Ra means the Flower of Knowledge. Same terminology applies to several Middle East languages, Western languages, and most particularly to Latin and French. And all have the same meaning. Wardah means Rose in Arabic. In Hebrew, it is Vered, and a female name was constructed from it, Varda. In Europe, the organization of the Rosicrucians took its name from the Rose.

Annunaki's masters are given the degree '33' just as they do in Freemasonry where the highest level in 33. Incidentally, the number three is important to all these people. For example, in the ancient calendar of the Anunnaki-Phoenicians, the month was divided into three weeks. Each week was composed of eleven days. The month consisted of three weeks, totaling thirty-three days."

"How fascinating," I said. "A few minutes into the Academy and I have already learned more about the truth than in months at the University on earth."

"It's not the professors' fault," said the apprentice. "Their knowledge is limited because so much has been done to obscure the truth, by so many different authorities. Sometimes I marvel how much the earth scholars did manage to learn, despite all the obstacles. But we must come in now, you have your purification ritual to do."

"Purification ritual? Why?" I asked.

---

[4] A Phoenician god, also worshipped by other people of the region

"All Anunnaki students are required to purify their bodies before their orientation or their regular course of studies. Lots of people do. You might remember that the Jewish scribes, since earliest times, had to purify their body before adding the name of God into the Torah? It is the same principle. The only difference is that our purification does more than purify the body. The substance we use purifies the mind and spirit as well. It's all very pleasant."

"Is it like the Jewish Mikvah? They have these communal baths that women visit, much like a swimming pool, where they purify themselves each month."

"No, ours is private, each student must purify his or her own being alone, since if we did it together, the mixing of the impurities might produce a barrier to the proper purification. Incidentally, remember the Essenes, these Judaic sect members of the Second Temple era? At first, they used our style of purification, but as time went by, and their numbers grew, they changed into a Mikvah-like, communal purification."

"What about the Christian baptism?"

"Of course, it is all the same idea of purity and cleanliness. And the Christians believe that the mind and spirit are indeed cleansed by the baptism."

He lead me to a door at the end of the room and opened it. Inside was a small room, entirely made of shimmering white marble. In the middle of the room was a basin, made of the same material, and filled with something I could not define.

"It does not look like water," I said, eyeing the glowing substance with suspicion.

"No, it is something else. This is Nou-Rah Shams, an electro-plasma substance that appears like 'liquid-light.' It actually means, in Anakh, The Liquid of Light. Nou, or Nour, or sometimes Menour, or Menou-Ra, means light. Shams means sun. Nour in Arabic means light.

The Ulema in Egypt, Syria, Iraq and Lebanon use the same word in their opening ceremony. Sometimes, the word Nour becomes Nar, which means fire. This is intentional, because the Ulemas, like the Phoenicians, believed in fire as a symbolic procedure for the purification of thoughts. This created the word Min-Nawar, meaning the enlightened or surrounded with light. In Anakh it is: Menour-Rah.

Which, if you know any Hebrew, you might remember that Menorah means a lamp.

It's all connected. Later, the Illuminati used it as well. Anyway, to complete the first phase, all you have to do is spend a little time in the basin and enjoy the purification. I will knock on the door shortly."

The Nif-Malka-Roo'h-Dosh Ritual

He closed the door gently behind him, and I removed my clothes and entered the basin. This bath was the most pleasant cleaning experience I have ever encountered. Every minute made me feel lighter, happier, more complete within myself and sparkling clean. I was sorry when the apprentice knocked on the door and asked me to get dressed and join him.

"Now," said the apprentice, "You are ready for the second phase of the purification. It is called the Nif-Malka-Roo'h-Dosh" Ritual"

"Heavens, what a name," I said. "What does it mean?"

"Nif means mind. Centuries later, it was used by terrestrials as Nifs or Roo'h, meaning Soul or Spirit. Since the Anunnaki did not believe in a 'separate soul,' the mind was the only source of creation and mental development, while humans continued to interpret it as 'Soul.' It means the same thing in Akkadian, Hittite, Aramaic, Hebrew and Arabic.

Malka means kingdom or a higher level of knowledge and mental development. Humans changed Malka to Malakoot or Malkout; and the same, or very similar word, was again used in Aramaic, Hebrew, Syriac, Coptic, Arabic, Phoenician and so many other languages.

Roo'h is the highest level of mental achievement. The Arabs use the same word, while the Hebrew word is Ru'ach. However, the meaning changed in both languages, to represent soul, not mind. And Dosh means revered. Now, let's go into the room where we can proceed." He opened another door, and we entered a moderately sized room off the classroom. In the room was a cell, shaped as a cone and transparent. The cell floated in the air, approximately twelve centimeters above the ground.[5] The top of the cell seemed to be connected to a beam originating from a grid attached to the ceiling, but the interesting thing was that

---

[5] Approximately 4.7 inches

the ceiling, too, floated in the air. It was totally suspended on its own.

"Please step into the cone," said the apprentice. A door opened and I entered the contraption, much intrigued. I stood and waited, and suddenly, a clear fog formed in the center. After a short while, the fog's color changed from white to silver-blue in form of waves.

I felt nothing at all, but to my amazement, I saw something registering on an information board that was posted on the right side of the cell. I looked at it, bewildered, and suddenly I knew, with absolute certainty, that these were my thoughts that were registering on the machine.

These thoughts began to take physical shape, which was instantly copied to a screen.

The screen transformed the thoughts-form into a code, which I could not decipher, but very quickly the code was transformed into a sequence of numerical values. I was not sure what this mysterious sequence meant, but later I found out that the sequence then is prescribed as a Genetic formula.

This genetic formula is the "Identity Registration" of the Anunnaki student. In terrestrial terms, you can call it DNA. But it is more than that. It is the level of mental readiness for the next stage.

At this point, I heard a direction in my head, as clear as if someone was talking to me directly. Obviously, I was approached on a telepathic level, something I was not as yet too accustomed to.

I was told to free my mind from all thoughts. It is something like what the Japanese call "Koan," or "Kara," a state of "mind nothingness." Surprisingly, I managed to do so with great ease.

I imagined this had something to do with the purification rituals, because in the past, when I tried to do the same in order to meditate, I failed miserably. Then, something began to happen.

Rays of various densities and colors surrounded me in a cloud. It is tempting to compare it to the aura in terrestrial terms, but this is not the case. It is not an aura, because it is not bio-organic. It is entirely mental.

*** *** ***

The creation of the mental Conduit
The "Double" and "Other Copy" of the Mind and body of the student

What happened next took only one minute, and at the time I could not understand what was really happening, but later it was made extremely clear.

> This was the most important procedure done for each Anunnaki student on the first day of his or her studies – the creation of the mental Conduit. A new identity is created for each Anunnaki student by the development of a new pathway in his or her mind, connecting the student to the rest of the Anunnaki's psyche.

Simultaneously, the cells check with the "other copy" of the mind and body of the Anunnaki student, to make sure that the "Double" and "Other Copy" of the mind and body of the student are totally clean.

During this phase, the Anunnaki student temporarily loses his or her memory, for a very short time.

This is how the telepathic faculty is developed, or enhanced in everyone.

It is necessary, since to serve the total community of the Anunnaki, the individual program inside each Anunnaki student is immediately shared with everybody. Incidentally, this is why there is such a big difference between extra-terrestrial and human telepathy.

On earth, no one ever succeeds in emptying the whole metal content from human cells like the Anunnaki are so adept in doing, and the Conduit cannot be formed.

Lacking the Conduit that is built for each Anunnaki, the human mind is not capable of communication with the extra-terrestrials.

However, don't think for a moment that there is any kind of invasion of privacy.

The simplistic idea of any of your friends tapping into your private thoughts does not exist for the Anunnaki. Their telepathy is rather complicated.

The Anunnaki have collective intelligence and individual intelligence

> The Anunnaki have collective intelligence and individual intelligence. And this is directly connected to two things: the first is the access to the "Community Depot of Knowledge" that any Anunnaki can tap in and update or acquire additional knowledge. The second is an "individual Prevention Shield," also referred to as "Personal Privacy." This means that an Anunnaki can switch on and off their direct link, or perhaps better defined as a channel, to other Anunnaki.

By establishing the "Screen" or "Filter" an Anunnaki can block others from either communication with him or her, or simply prevent others from reading any personal thought. "Filter" "Screen" and "Shield" are interchangeably used to describe the privacy protection.

In addition, an Anunnaki can program telepathy and set it up on chosen channels, exactly as we turn on our radio set and select the station we wish to listen to. Telepathy has several frequency, channels and stations.

When the establishment of the Conduit is complete, the student leaves the conic cell and heads toward the section assigned to him or her at the classroom. In my case, I was just about to finish my orientation with the all important conversation with Sinhar Inannaschamra, discuss my mission with her before it was to be communicated to the Council, and make a few important decisions for the future, both personal and professional.

*** *** ***

Chapter 57
On extraterrestrials currently living among humans and USO

- Categories of extraterrestrials living among us...155
- Extraterrestrial Grays from various stars and planets...155
- Hybrids who are half humans-half aliens...155
- Extraterrestrials from various spatial races who live underwater...155
- The USO...156
- Many reports were submitted by sea captains and authenticated by military crews...156
- The recorded USOs' reports are in the thousands...157
- The "Malta Water-UFO/USO...157
- The "American UFO/USO-spy plane and metal memory substance...157
- The Argentinian 500 miles per hour USO...158
- Lake District, United Kingdom's USO sightings...159
- Laguna Cartegena USO activities...159
- But why underwater bases? ...159
- Ideal locations for extraterrestrials...159
- Are they harmful to us? ...160
- They are not, because they are from our future...160
- Extraterrestrials cannot alter the past...160
- Essential and unique information about USO...160

*** *** ***

Chapter 57
On extraterrestrials currently living among humans and USO

Categories of extraterrestrials living among us

They are not all over the world as many tend to believe. The majority of extraterrestrials live in the United States, on land, underground and underwater. They are of 3 categories:

1-Extraterrestrial Grays from various stars and planets:

They came from various stars and planets, from our galaxy and far distant ones. They are not exclusivel the Grays from Zeta Reticuli, as many believe. They work in secret military bases, installations and laboratories.
They work together with American scientists on various projects ranging from quantum physics, genetic research/experiments and mind control to various paranormal programs, weapons system development, special spaceships capable of traveling to areas and spatial regions beyond the solar system, and on global intergalactic communications.
Many of them live in separate floors in underground bases. They do not show themselves to the general population.

*** *** ***

2-Hybrids who are half humans-half aliens:

Many hybrids live in the United States with American families who have adopted them. Usually high level military personel families.
Other hybrids live in horrible habitats in restricted areas under the control of the Grays. They live a miserable life.

*** *** ***

3-Extraterrestrials from various spatial races who live underwater:

They have a well developed habitat. Many of these habitats are in the area of Alaska, the Pacific and Puerto Rico. These races are the ones who abduct human for a multitude of reasons, including genetic experiments and reproduction.

They are not very friendly toward the United States Government. The American military and top echelon scientists working on "Black Projects" feel that they are a real threat to national security.

These extraterrestrials are extremely advanced, and they have mastered the sub and supersonic naval navigation systems. Their spaceships are called USO.

The USO
Many reports were submitted by sea captains and authenticated by military crews

I spent 20 years researching and studying the USO phenomena. In addition, I interviewed more than 100 persons from all walks of life who assured me that they saw flying machines coming out of the sea. Among the individuals I talked to were ambassadors, admirals, navigators, commercial airlines and military pilots, topographers, explorers, scientists, marine biologists, and of course some visionaries and "people who see things!" Also, I was fortunate enough to look into logs and reports submitted to the US Navy, the British Admiralty, French secret service files, and several other naval authorities in Spain, Brazil, Cyprus, Lebanon and Puerto Rico. Many reports were submitted by sea captains and authenticated by military crews. Yet, there is no single, physical and authoritative evidence, ever produced by an official authority!

I can easily understand the reasons.

The USO's subject is fascinating and remarkably intriguing because many scientists-ufologists are fully convinced that the unidentified submerged objects (USO) are *de facto,* the "real" UFOs we see in our skies. "They just come out of the sea, fly over...do what they have to do...and return back to their underwater bases...I saw 2 water-UFO myself...they had bright lights, circular...and flew very slowly at the beginning...and then all of a sudden they flew at un unbelievable speed...I reported the first sighting to my superior...and you know what he said to me...You still have problems with Sylvia, Eh?...Sylvia is my wife..." told me once, a highly ranked naval officer in Brazil.

*** *** ***

The recorded USOs' reports are in the thousands

The most notable ones are:

1-The "Malta Water-UFO/USO": In 1947, an immense USO (crescent shape) was seen in the Mediterranean sea, and frogmen climbed aboard. High ranking officers at the British Admirality investigated the incident, and concluded that the "flying object" was nothing but a submarine. I am happy they did not say it was a "sea-air baloon."
Two months later, the same or identical USO-UFO flew over Malta for a very short time, and according to several fishermen, the USO "entered the water at a great speed without producing water splashes, as if you were cutting a piece of butter with a butcher's knive..." This time, no one at the Admiralty would dare to say it was a flying submarine.
The USO's sighting is still well-remembered and talked about in the island of Malta.
The sightings of UFOs traversing the ocean are not new.
On June 18, 1845 the Malta Times wrote: "We find the brigantine Victoria some 900 miles east of Adalia, when her crew saw three luminous bodies emerge from the sea into the air. They were visible for ten minutes, flying a half mile from the ship." There were other witnesses who saw the same UFO phenomena from Adalia, Syria and Malta. The luminous bodies each displayed an apparent diameter larger than the size of the full moon.

2-The "American UFO/USO-Spy Plane and Metal Memory Substance": In August 1958, two UFOs-USOs were spotted above the sea level in "Al Manara" area of Beirut, Lebanon. People thought they were American "spy-planes."

Lebanese communists reported the sighting to the Soviet Embassy, located 2 miles from the so-called "USO hot spot."

In September, the USOs resurfaced exactly in the same location, but this time they shot 2 red beams over the shore, and a strange black substance fell off from under the belly of the USOs. For some mysterious reasons, this "substance" ended in the hands of biologists at the AUB (American University of Beirut.)
In 1958, the first civil war in Lebanon exploded, dividing the country between Muslims backed by Syria and Egypt's Gamal Abdel Nasser, and Christians who hoped that the Americans will come to their rescue.
Dwight Eisenhower sent the 6th fleet to Lebanon upon the urgent request of then President Camile Chamoun.
Everbody in the region believed that USOs were American military planes launched from the 6th fleet, except the Russians who knew better.
Years later, upon visiting a friend, Mr. Pierre Millet, the former French Ambassador to Lebanon, and talking about all sorts of things ranging from the Algerian crisis to Boileau and Voltaire, the subject of UFO was brought up. Mr. Millet told me verbatim: "Those things exist…and the Americans know a lot about UFOs…I have seen a report that detailed exactly how these flying saucers (Soucoupes Volantes in French) work…there is also another kind that comes out of the sea…these "Soucoups Volantes" are the ones that map the earth and spy on military installations…"
I asked him about the mysterious "black substance" recovered by the Americans, and Mr. Millet said: "Ah oui…it is true…the Americans got it…I think the American University in Beirut got it…it was a "chunk" and it was sent to an American military headquarter in West Germany…they found out that the substance had some sort of a metal memory, and it was made out of a very strange gaz composition and coagulated combustion molecules, it was not liquid…strange…very unreal…this was quite revolutionary and unknown in those days…"

3-"The Argentinian 500 Miles Per Hour USO": In January 1960 in Golfo Nuevo, a huge unidentified submarine was spotted in the Argentinian waters, and was hunted down by 2 Argentinian cruisers and one destroyer, joined by 35 navy planes. The submarine escaped the chase at an enormous speed exceeding 500 miles per hour. At that time, and with a mediocre data, military observers and civilians as well thought it as a secret German weapon remnant. Others believed that it was the latest submarine type 21 developed by the Nazis during World War Two.

4-Lake District, United Kingdom USO Sightings: Since the late 1980s, there have been numerous and continuous sightings of USO's around the Lake District, UK. The most famous one was the USO sighting in 1994 at Derwent Water. The sighting lasted almost five minutes and was observed by 25 people. The area now receives regular USO sightings reports, including from surrounding areas. The most recent UFO/USO sighting occured in January of 2006, and the most recent USO sighting was reported in December 2004.

5-Laguna Cartegena UFO/USO Activities: The area surrounding Laguna Cartegena in Puerto Rico is a location considered to be the earth's USO headquarters with intense USO activities. This is the USO hottest spot, which sightings reports frequently originate from.

*** *** ***

But why underwater bases?
Ideal locations for extraterrestrials

Geologists in the East and West coasts are busy understanding a new theory that shows possible underground UFO bases all around the world. According to this theory, the UFO bases are along the interface of the seven large and many small tectonic plates meet each other.
According to UFO researchers, the underground UFO bases are deep under the ground where multiple tectonic plates push on top of each other. For example the Indian plate and Eurasian plates colliding against each other along the Himalayas makes it ideal locations.
According to this theory, the UFO bases need to be deep under the ground because the UFO crafts need to be close to the mantle

of the earth. Servicing of these crafts can be done in that electromagnetic environment only.

In addition according to this theory the crust must be as thick as possible in that area. That is only available where one tectonic plate moves on top of another tectonic plate. (Source: India Daily.)

Are they harmful to us?
They are not, because they are from our future

Many ufologists believe that USOs' aliens abduct people, the same way UFOs" aliens do. Although, fewer alien abductions were reported, the number is not negligible, and those incidents deserve wider attention.

Another group of ufologists remains optimistic, for the aliens are here on earth to help humanity. But the piece de resistance comes from those who advocate the theory that UFOs-USOs are no threat to us, because they are from another time, and another dimension: Our future.

*** *** ***

Extraterrestrials cannot alter the past

According to some theories, extraterrestrials cannot alter the past. That means they cannot be harmful to us. They are also of no benefit to us.

The reasoning behind this theory is that:
- In the parallel dimension, the spatial structures can be viewed, but like in physical universe there is nothing to alter.
- Once in space-time equations, space is fixed and time is accelerated back wards, the space cannot be altered.

*** *** **

Essential and unique information about USO

Victoria (Sinhar Ambar Anati), the hybrid Anunnaki-human wife of Sinhar Marduch, an Anunnaki leader, knows a lot about aliens' underwater habitat and bases. Victoria has visited few extraterrestrials' underwater bases, and talked to aliens in their native languages.

In my rapport with her, Victoria explained to me the reasons and purposes of the aliens' underwater bases, and described in detail their families, way of life, technology, and especially their relationship with top echelon United States military scientists. Herewith are excerpts from her revelations:

- 1-Water UFOs are the sky and land UFOs. The UFOs you see in your skies come from bases underwater on your planet.

- 2-On earth, it is extremely rare to sight UFOs coming from outer space, from and beyond your galaxy. It did happen a few times, but what you saw were machines from your future.

- 3-Your past (in your terrestrial language) is relatively young. Your future is a distant past in the universe.

- 4-Events that are currently happening on other planets will not be seen or understood by humans before millions of years.

- 5-Water UFOs are ageless. They were manufactured millions of years in your past. They do not suffer from metal fatigue, because the metal used has its own memory. And that memory reconditions metals.

- 6-Underwater, the color of the USOs changes according to the environment, the density of the water and the aquatic conditions of the milieu.

- 7-Many USOs are surrounded by an anti-gravity perimeter and protected by a shield non visible to the naked eye and to the most sophisticated radar on earth, or any observation/reconnaissance device. Don't try to understand it, it is way beyond the intellect and intelligence of human beings.

- 8-The most advanced type of USOs are the crescent-shaped ones. They are smaller than the circular ones, and usually are deployed over mountainous areas.

- 9-They fly in formations, and are tele-guided by a space-time memory apparatus. It works like a navigation device, but not like a compass, because it does not have altitude, longitude and directions/positions of your south, north, west and east.

- 10-There are no linear or trajectories information on board of the USOs, because, in reality, USOs-UFOs do not fly; they change positions, in other words, they "jump from one pocket to another."

- 11-Pocket means the "rendez-vous" of entry and exist of "anything" in the time-space opening that occurs when an object, even a thought escapes the "measurable." I told you, it is very confusing and seems unrealistic to you. But to millions of civilizations in the universe, it is a basic scientific knowledge.

- 12-On earth, scientists tried to explain this "rendez-vous" of time and space as "parallel universes", "multiple universes", "higher dimensions, "the M Theory", "future universes", "wormholes", "time warp", "black holes attracting forces", "reversed time", "time travel", "space-time travel", "black holes absorbing energy". This is the best they could do. I must admit they were very prolific.

- 13-The "rendez-vous" is infinitesimally small universe molecule capable of expanding into the shape of a tube, where physical energy and earthy laws of physics cease to exist; a UFO can penetrate this molecule (call it tube if you want or even wormhole) and reach multiple destinations faster than a beam of light. And the opening of the "rendez-vous" exists only for a fraction of time. I used "time" because that is the only measurement you can understand for now.

- 14-The speed of light at that fraction of time is created by a negative energy that distorts the fabric of space; hence, time as you know it and understand it ceases to exist.

- 15-Extraterrestrials (as you call them) are capable of opening this "rendez-vous", travel through it, and exit before it closes up or collapses on its own time-space dimensions.

- 16--USOs penetrate the water without creating splashes, and come out of the water "without dripping."

- 17-Before the USO touches the surface of the water, it shoots an invisible beam that separates the lower part of the space ship from the water, thus creating a "vacuum space" similar to an air pocket.

- 18-USOs do not navigate waters. They "slide" through an "air-vacuumed tunnel" that opens up right in front of them, and closes behind them as soon as they cross it. The water never touches the body of the scapeship.

- 19-When a so-called abductee begins to believe or tell you that he/she is being chosen as aliens' messenger to humanity; it is time to rethink the whole abduction's scenario.

- 20-The "original Anunnaki" no longer live on earth. They left your planet thousands of years ago. However, many of their off springs, bloodlines, descendants and hybrid remnants live among you.

- 21-The extraterrestrials (as you call them) of the sea are not Anunnaki. They belong to a different race; one of the 46 different alien races that have visited the earth at its dawn.

- 22-Only 7 specific races remained on earth. They have many things in common, and share non-physical similarities, but are different from the Anunnaki. The aliens who currently work with terrestrial scientists have "extraterrestrial physiognomy", totally alien to the human race.

- 23-Anunnaki are not very much different from human beings. And they are not reptilians like some of the races that inhabit Zeta Reticuli.

*** *** ***

Chapter 58
Key words, important terminologies and expressions from the Anunnaki language and ancient Middle Eastern epics/tablets...261

Excerpts from Maximillien de Lafayette's book:
Thesaurus-Dictionary Of Sumerian Anunnaki Babylonian Mesopotamian Assyrian Akkadian Aramaic Hittite: World's first languages & civilizations: Terminology & relation to history, Ulema & extraterrestrials."

- What or who is Aa?...169
- Who is Abd?...170
- The genetic composition of Abel...170
- Eve's relation to the Anunnaki...171
- The concept of Eve conceiving Cain and Abel with the help of God...171
- What is Abra-ah? ...172
- What is Afarit? ...172
- What is Amram? ...173
- Abraham, God and the Anunnaki face to face...173
- What is An'h-Ista-Khan-na-reh? ...174
- What is Anda-anparu? ...175
- What is An-Maha-Rit? ...175
- What is An-Nafar JinMarkah? ...176
- What is Anunnaki ME.nou-Ra "MEnour"? ...177
- What is Anunnaki.Nou-Rah-Shams."Menora-Shems"? ...177
- What is An-zalubirach? ...178
- What is Anzuma? ...178
- What is Arakh-nara "Arcturus"? ...178
- Edgar Cayce did mention it...179
- Arcturus is the first docking station that allows us to travel beyond our consciousness...179
- What is Arwad "Aradus"? ...179
- Arwad hold many great secrets, to name a few...180
- What is Ay'inbet? ...181
- Baalshalimroot...181

- Those who have practiced As Sihr (Magic) used their eyes as a psychic conduit...182
- The concept of hell was unknown to the pagan Badou Rouhal...183
- Esoteric benefits...183
- What is Azaridu? ...183
- A selection of important words from the Anunnaki's language...184
- Aamala...184
- Abamarash...184
- Abekira'h-Kitbu...184
- Aberuchimiti...184
- Abgaru...185
- Abra-ah...185
- Akama-ra...185
- Anšeduba...185
- Anubdada...186
- An-zalubirach...187
- Anzuma...187
- Baliba nahr usu na Ram...188
- Balu-ram-haba...188
- Barak yom-ur...190
- Barak-malku...190
- Chavad-nitrin...191
- Chimiti...191
- Dab'Laa...191
- Dadmim "Admi", "Adamai", "Adami" ...192
- Da-Irat...192
- Dayyakura...197
- Dhikuru...197
- Eido-Rah...197
- Emim...198
- Ezakarerdi, "E-zakar-erdi "Azakar.Ki" ...199
- Ezakarfalki "E-zakar-falki"...200
- Ezbahaiim-erdi...200
- Ezrai-il...200
- F "ف" ...201
- Fari-narif "Fari-Hanif" ...201
- Ghen-ardi-vardeh "Gen-adi-warkah" ...201

- Gens "Jenesh" ...205
- Gensi-uzuru...205
- Ghoolim...206
- Giabiru...207
- Gibbori...207
- Gibishi...207
- Gibsut-sar...207
- Gilgoolim...208
- Ginidu...208
- Girzutil...208
- Goirim-daru...208
- Golibu "Golibri" ...208
- Golim...208
- Gomari "Gumaridu...209
- Gomatirach-minzari "Gomu- minzaar"...213
- Prerequisites...215
- Precautions...216
- Equipment and supplies...217
- Building the Minzar...218
- Contacting the alternate realities...219
- Subsequent visits to the alternate realities...222
- Benefits and advantages...222
- Returning to your regular reality on Earth...223
- Gubada-Ari...224
- Synopsis of the theory...225
- Materials...225
- The technique...226
- Gudinh...232
- Hag-Addar...232
- Hamnika-mekhakeh...232
- Hamnika-mekhakeh- ilmu...232
- Synopsis of the concept...232
- The Ulema-Anunnaki days are...232
- The calendars' grids...233
- The use of a language...233
- The preparation and use of the grids...234
- Grid 1: Calendar of the week...234
- Grid 2: Calendar of your name...235
- Calendar of your lucky hour...236

- Grid 4...236
- Grid 5...237
- Hamsha-uduri "Dudurisar" ...238
- The concept...238
- The technique: It works like this...239
- How real is the holographic/parallel dimension you are visiting? ...245
- Some of the benefits...247
- Closing the technique...247
- Harabah...248
- Haridu "Haridu-ilmu" ...248
- Harima...251
- Harimu...251
- Harranur-urdi...251
- Functions of the Conduit, Miraya and retrieving data...254
- Hatani...262
- The concept...262
- The Hatani protection shield...264
- Hatori "Hatori-shabah" ...266
- Hattari...267
- Hawaa...289
- Hawwah...289
- Hawwah/Eve: Transliteration of a text from the Book of Ramadosh...291
- Translation of the text from the Book of Ramadosh...292
- Haya-Saadiraat...293
- Hazi-minzar "Mnaizar" ...293

*** *** ***

Chapter 58
Keywords, important terminologies and expressions from the Anunnaki language and ancient Middle Eastern epics/tablets

What or who is Aa?

The first meaning:
Aa "Ai" is an Assyrian expression meaning the female strength of the sun. Usually written as Aa na shams. Shams means sun in Assyrian; in Arabic, it is Shams; In Hebrew, it is Shemesh. In Phoenician, it is Shama or Shem. When Aa is used as Ai, the meaning becomes: Negative; enemy.
Nebuchadnezzar said: "Ai isi nakiri"; meaning: May I not have enemies.
Sardanapalus said: "Kasid ai-but Assur," meaning: Capturing the enemies of Assur.
In the Anunnaki's literature, women played a major role in the human affairs, as well as in the first three Anunnaki's expeditions to planet earth.
Anunnaki's women or goddesses were the creators of the human race; they were the first geneticists who produced the seven human prototypes, and adjusted the Conduit in the brains' cells of the early humans. (See Creation: Seven prototypes).
Anunnaki's goddesses were known to the early Phoenicians of Arwad as the sun-goddesses.
They had both positive and negative energies, depending on the intentions of each goddess. The Assyrian concept of the female strength of the sun derived from local Phoenician legends, based on oral history and tales known to the Arwadians.

The second meaning:
Aa is also the Assyrian, Akkadian, Babylonian, Sumerian, and Mesopotanian name of the Babylonian, Sumerian and Assyrian deity often referred to as Aê, and Ea. He is also represented by and identified as Au, Ya'u and Ya which is a variation of Ea, an

ancient Babylonian deity. Ya corresponds to the Hebrew Au, Aw, Awu. From Ya, the Hebrew Yah or Jah were derived, and used as prefix for Yahweh.
Originally derived from the Ana'kh Aa'h, which means leader or creator.
Aa had numerous names and titles; he was the Babylonian and Assyrian god of water, rivers, the sea, the arts, and crafts. He warned Pir-napistim of the Deluge, and instructed him to build a ship to save his family, himself, all the birds, and the animals of the earth.
Worth mentioning that the Babylonian Pir-napistim became the Chaldean/Biblical Noah.

Who is Abd?
Abd is a Sumerian, Anak'h, Arabic, Babylonian, and Ulemite noun. Historically, it was the first name given to Man. The original meaning was slave, but later on, Enki changed it to servant.
In contemporary Arabic, it is written either as Abd or Abed and it means two things:

1-A black person;
2: A slave.
Many derogatory attributions for Abd are found in the Arabic poems of Abu Al Tib Al Mutanabbi (915-65 A.D.), in the writings of Abu Al Ala' Al Maari (died in 1057), and Al-Nabigha Al-Zoubyani (535-604), and especially in the story of king Dabshalim and Brahman Baydaba. (Around 175 B.C.)

The genetic composition of Abel
In order to explain and understand the genetic composition of Abel, an Ulema suggested that we should ask ourselves what kind of relation Eve and her children had with God and the Anunnaki. There is a vast literature about Eve, and lots of contradictory accounts about her true nature, her origin, her DNA, and above all, her relation with the Anunnaki, the Gods, and the Judeo-Christian-Muslim God. Eve appeared in the Sumerian texts, in Phoenicians epics, in the Bible, in the Quran, in the Gnostics books, and in the Ulema's manuscripts. Eve story in the Bible is the less credible one.

Eve's relation to the Anunnaki

Humans who were genetically created by the Anunnaki were produced from and by a mixture of the DNA of an Anunnaki, usually a god or a goddess, and an earthy element. This element was described as either clay or specie of a primitive human being. The intervention of an Anunnaki god was a prerequisite.
Thousands of years later, the Bible told us that Eve too received a divine help in the creation of her first two sons; they were fathered by the Lord not by Adam.

The concept of Eve conceiving Cain and Abel with the help of God

This could and would astonish the Christians. Eve conceived Cain and Abel with the help of God. Only her third son Seth was the result of her union with Adam. And Seth came to life in Adam's likeness. So how did Cain and Abel look like?
The Bible does not provide an answer. From Genesis: 4:1 "...and she bore Cain saying: I have gotten a man with the help of the Lord. And again, she bore his brother Abel..."
Genesis 5:3: "When Adam had lived a hundred and thirty years, he became the father of a son in his own likeness, after his image, and named him Seth."

The Gnostics books shed a bright light on this situation; Cain was created by the Anunnaki god En.KI and a woman called KaVa, (Also Havvah and Hawwa) which is the original name of Eve in the ancient texts written thousands of years before the Bible was written and assembled. This is the official version of the Gnostics.
This means that Cain is not 100% human. Cain's blood is ¾ or ½ Anunnaki. The other two sons of Eve, Abel called "Hevel", and Seth called "Sata-Na-il" were less than ½ genetically Anunnaki, because they were the offspring of KaVa (Eve) and Ata.Bba (Original name of Adam). Cain was superior to his brother Abel at so many levels, because he was the offspring of

an Anunnaki. Abel was inferior to Cain, because he was the offspring of an earthy element. The superiority of Cain was documented in the Bible, because the Bible (Old and New Testaments) clearly stated that Cain "rose far above Abel"!
Thus, the Ulema, conclude that:

- **1-**Eve and Adam were not from the same race. Genetically, they were different.
- **2-** The offspring people (First human race) of Eve were the result of a breeding by Gods.
- **3-**The children of Abel and Cain were genetically modified to fit the scenario of the Anunnaki.
- **4-**The creation of the human race happened earlier, much earlier than the date suggested by Jewish, Christian and Muslim scriptures.
- **5-**All human races came from the primordial female element: Eve.

What is Abra-ah?
Abra-ah is an Ana'kh noun. It is the name of the fifth element in the Anunnaki's matrix. It means transcending time and space.
The Ulema coined it "Niktat Alkhou-Lood", and it means verbatim: The point of the beginning of immortality. In other words, that fifth element is responsible for an extraordinary longevity of mankind on earth, and/or its immortality. It is very possible said an Ulema that after 2022, humans will learn about the secret of immortality, but will never be able to decode the composition and sequences of the fifth element. It would be a catastrophe for humanity and for the future of planet earth, if humans succeed in decoding the data contained in the fifth element. Many Ulema are not seriously worried, because with the arrival of the Anunnaki in 2022, the existence of human life and its continuity will be in the hands of the Anunnaki.

What is Afarit?

Afarit is an Arabic noun, derived from the Ana'kh Afa-rit, meaning the entities with extraordinary powers. Afarit are usually associated with talismans.

> A Talisman is a small amulet or other object, often bearing magical symbols, worn for protection against evil spirits or the supernatural. Demons are ordered in talismans to follow the instructions and to leave the patient whom they inhabit. A spirit of the lower world is assigned to every day of the week.

Some of the most important names are:

- El Mudhib, known as abu 'Abdallah Sa'id rules over Sunday;
- Murrah el-Abiad abu el-Hareth (Abu n-Nur) over Monday;
- Abu Mihriz (or abu Ya'qub) El Ahmar rules over Tuesday;
- Barqan abu l-'Adja'yb rules over Wednesday;
- Shamhurish (el-Tayyar) rules over Thursday;
- Abu Hasan Zoba'Ah (el-Abiad) rules over Friday;
- Abu Nuh Meimun rules over Saturday.

What is Amram?

Abraham, God and the Anunnaki face to face

Amram is an Ana'kh noun. and means the good subjects of the Anunnaki's leaders, and/or good communities. It is composed of two words:

- 1-Am (Good; kind)
- 2-Ram (People; community; population; tribe.)

In Biblical studies, Amram means high people; kindred of the High; friend of Jehovah. In primitive Arabic, Ram meant: People; group. Henceforth, the name of the Palestinian city Ramallah could be interpreted as the people of God, since Allah means to the Arabs and Muslims, what exactly the word Jehovah means to the Jews: God. When Enki or Ea called upon Avraham, he told him: I am your god, and I am now changing your name

from Av-raham to Ab-Raham, because you are going to lead my people as the father of my people on earth. Av became Ab. And Ab in all the 14 different ancient languages of the Near East and the Middle East means father. From the Ana'kh Ab, derived the words: Ab, Abu, Abi, Aba, Abba, Abuya, Abouna; all meaning the very same thing: Father. And from the Ana'kh word ram, derived the ancient Hebrew, Aramaic and Arabic word Ram: People.
Centuries later, Ram acquired a multitude of meanings. For instance:

- **a**-In ancient Hebrew, Ram is pleasing;
- **b**-In Sanskrit mythology, Ram means supreme.
- **c**-In the pre-Islamic Arabic era, called Al-Jahiliya, (Ages of darkness), Ram meant a group of people. Synonym: Ra'bh.

What is An'h-Ista-Khan-na-reh?
An'h-Ista-Khan-na-reh is an Anunnaki word for women's impregnation operation in Nibiru. The woman goes into a very nice hospital-like place. Anunnaki physicians will help her to lie down on a table, much like one of the examination tables in any doctor's office on earth.
The attending physicians will be all females, very gentle and extremely skilled. Using a special machine, they will beam a light right through the woman's body; the light will search for her ovaries.
Nothing will probe, or hurt, or even annoy the body. Once the light reaches the ovaries, it will activate one of the eggs, fertilize it, and have it move very smoothly into the uterus. The woman then becomes pregnant, and the fetus will begin to grow. Anunnaki women have the egg removed by the light, placed in a special tube, and grow the baby in a machine.
They don't have birth in the same sense humans do, but take the baby home after he or she is ready in the advanced incubator. This impregnation operation is called "An-Ista-Khan-na-reh".
It means the following:

- **1**-An'h=Creation, first, celestial;
- **2**-Ista or Ishtah=Child, baby, first born;
- **3**-Khan=hospital, operating room;

- 4-Na=Source, first breath, first nourishment;
- 5-Reh or Rah=Delivery, reception, relief.

What is Anda-anparu?

> You are always connected to the Anunnaki in this life and the next one

Anda-anparu is an Ana'kh/Ulemite expression meaning verbatim: Judgment by the brain. Ulema Saddik Ghandar Ranpour said: "Seconds before you leave earth, your mind will project the reenactment of all the events and acts (Bad and good) in your life, past, present and future, and "zoom" you right toward your next nonphysical destination, where and when you judge yourself, your deeds, your existence and decide whether you wish to elevate yourself to a higher dimension, or stay in the state of nothingness and loneliness.

You will not return to earth, nor will your soul migrate to another soul or another body, because the Anunnaki do not believe in reincarnation or in a physical return to earth. Earth is the lowest sphere of existence for humans. Thus, you are always connected to the Anunnaki in this life and the next one. The Anunnaki were the ones who created the brain for the humans. These early brains contained two million cells. But the Anunnaki worked years on the prototypes of humans. In their final genetic experiments, the Anunnaki programmed humans with the thirteen original faculties, implanted in the brain's cells."

What is An-Maha-Rit?
An-Maha-Rit is the Ana'kh, Phoenician, and Ulemite name for an Anunnaki's quasi formula for steroids. It is composed from three words:

- 1-An (First; sky; heaven; god)
- 2-Maha (Ultimate; great)
- 3-Rit (Strength; energy; motion)

Mah-Rit is humanity's early form/formula of what we call today steroids; an early genetic product created by the Anunnaki in the ancient Phoenician city of Amrit, in Syria co-built by the remnants of the Anunnaki and early Phoenicians from Tyre and Sidon.

Amrit is one of the most puzzling, mysterious and enigmatic cities in recorded history. It was the stage for a cosmic war between many ancient nations; the birth of the Olympiads; and the world's first Anunnaki-Phoenician medical center. But recent archeological excavations on the Island of Arwad revealed that this island gave birth to the Olympiads, and not Amrit as it was suggested by historians.

An-Mah-Rit was first used by Inanna when she created the first 7 prototypes of the human race. Phoenicians used An-Mah-Rit quite often. It was supplied by the priests of god Melkart.

What is An-Nafar JinMarkah?

An-Nafar JinMarkah is an Anak'h noun/expression. Nafar Jinmarkah, is the name of humans who walked on three legs, supposedly created by non-terrestrial geneticists who have visited earth some 450,000 years ago.
Some ufologists suggest that they were genetically created by the Igigi.
Some authors have claimed that the Igigi have created a very primitive form of human beings, lacing intelligence and mobility.

The early forms of humans looked like apes

The Igigi considered the early quasi-humans to be not much more than machines with limited mental faculties, and those early forms of humans looked like apes.

The earth was extremely cold at that time, and the Igigi had to cover the human bodies with lots of hair to protect them from the elements. It took the quasi-human race thousands of years to evolve into an early human form, and even then not totally human, still looking like apes. Some of them had bizarre skulls and facial bones. The Igigi actually experimented a bit with the early human-forms. First, they created the "Nafar Jinmarkah" meaning 'individual on three legs.' They consisted of a very strong physical body but lacked agility. Those bodies were created to carry heavy weight. The three legs' purpose was to support heavy loads they could lift and carry. Later on, the Igigi worked on a new human form that consisted of a body with two legs, in order to bring speed and better agility. Yet, early humans remained terrifying, nothing like the Biblical descriptions. The Igigi tried several times. And each time, they faced a problem in designing the human skull. Early Igigi creators did not want to put brains in the skull so human-forms-bodies would not think. These early human-forms were the world's first robots.

What is Anunnaki ME.nou-Ra "MEnour"?

Anunnaki ME.nou-Ra "MEnour" is the Ana'kh name for a sort of a light (Plasma laser) used by the Anunnaki to purify the body and thoughts. All Anunnaki students entering the classroom in an Anunnaki academy must purify their bodies and minds. The purification exercise occurs inside a small room, entirely made of shimmering white marble. In the middle of the room, there is a basin, made of the same material, and filled with a substance called Nou-Rah Shams; an electro-plasma substance that appears like liquid-light.

It actually means, in Ana'kh, the Liquid of Light. Nou, or Nour, or sometimes Menour, or Menou-Ra, means light.
Shams means sun.
Nour in Arabic means light.
The Ulema in Egypt, Syria, Iraq and Lebanon use the same word in their opening ceremony. Sometimes, the word Nour becomes Nar, which means fire. This is intentional, because the Ulemas, like the Phoenicians, believed in fire as a symbolic procedure to purify the thoughts.

What is Anunnaki Nou-Rah-Shams "Menora-Shems"?

Nou-Rah-Shams "Menora-Shems" is an Anak'h, Hebrew, Phoenician, Arabic, Akkadian, and Aramaic term. It is the Anunnaki's Liquid-Light; an electro-plasma substance that appears like luminous watery substance. In Ana'kh, Menou-Ra actually means the following: Nou, or Nour, or sometimes Menour, representing light. It has the same meaning in many Semitic languages. The Ulema, like the Phoenicians, believed in fire as a symbolic procedure to purify the thoughts. This created the word Min-Nawar, meaning the enlightened or surrounded with light. If you know any Hebrew, you might remember that Menorah means a lamp. It's all connected. Later, the Illuminati used it as well.

What is An-zalubirach?

An-zalubirach is an Ana'kh/Ulemite term for collecting thoughts and multiple mental images; using mental energy to move or teleport things. This is one of the phases and practices of Tarkiz.
Tarkiz means deep mental or intellectual concentration that produces telekinesis and teleportation phenomena. Anunnaki's young students, apprentices, and novices learn this technique in Anunnaki schools in Ashtari.
Usually, they use their Conduit (Which is located in the brain's cells), and deep concentration on an object hidden behind a screen or a divider made from thin paper. Synchronizing the frequency of their Conduit and an absolute state of introspection, the Anunnaki student attempts to move the hidden object from one place to another without even touching it. In a more advanced stage, the Anunnaki student attempts to alter the properties of the object by lowering or increasing the frequencies and vibrations of the object.

What is Anzuma?
Anzuma is the Anak'h term for the Anunnaki's designated spots or areas of landing on earth. Usually, they are used to map terrains to serve as landing terminals. The first time the Anunnaki visited planet earth, they marked the regions where they have landed their Merkabah (Spaceships). These regions were the ancient Phoenicians cities of Tyre, Sidon Afka, Baalbeck (Baalback), Bijeh, Amchit, Byblos, Batroun and Arwad Island.

What is Arakh-nara "Arcturus"?
Arakh-nara "Arcturus" is the Ana'kh name of a planet, unknown to legitimate science. It is composed of two words:
- 1-Arakh (Portal or station)

- **2**-Nara (Star or light.)

It was inhabited by the Anunnaki some 700,000 years ago. Many of us are not familiar with. However, it was mentioned in channelers' séances and by Eastern ufologists, as well as by the Ulema.

Arakh-nara is known in the West as "Arcturus".

Edgar Cayce did mention it

He stated: "...one of the most advanced civilizations in this galaxy. It exists in fifth-dimension and is the prototype for arth's future. Its energies work as emotional, mental, and spiritual healers for humanity. The star is also an energy gateway through which humans pass during death and re-birth. It functions as a gateway station for non-physical consciousness to become accustomed to physicality. Arcturus is a stargate through which souls pass, to choose whether to return to the Earth-sun system, or evolve to others." J.J. Hurtak in his "Book of Knowledge: Keys of Enoch", wrote, "Arcturus is the midpoint between the Earth and other higher levels of consciousness. It is the governing body for our universe which determines the spiritual progress of humanity.

What is Arwad "Aradus"?

> Arwad "Aradus" is the Phoenician, Ugaritic, Greek, Arabic and Ana'kd name for an ancient Phoenician island in the Mediterranean Sea.
> The Island of Arwad was an independent kingdom in the days of the Canaanites. It was created by the Phoenicians in early second millennium B.C. This small beautiful island located 5 miles from the city of Tartus in Syria, was one of the first Anunnaki's small colonies on earth.

It was mentioned in the Bible by the Prophet Ezekiel. Arwad was the headquarters of the seven wise men who came from Apsu, the sweet water, and attended the gods of Enki. They were known to the Sumerians as Abgal, to the Akkadians as Akkallu, and to the Phoenicians as An-Khal. The Anunnaki called them "The Ab-n'GAL." On the Island of Arwad, the Phoenician created a secret society called the "Circle of the Serpent" to honor their god Melkart. On Arward, the Melkart shrine/altar still stands in all its beauty and majestic architecture. The early learned Greeks who visited Arwad studied medicine at the Phoenician-Anunnaki medical center, and when they returned home, they adopted the Phoenician sign of the serpent as the logo for their healing arts.

Arwad hold many great secrets, to name a few:

- **1**-For a short time, Jesus and Mary Magdalene lived there after the Biblical Crucifixion.
- **2**-St. Paul sailed to Arwad after he has spent some time in Byblos (Jbeil) and Batroon in Phoenicia (Today, Lebanon).
- **3**-It was at Arwad, that the Anunnaki created the "Brotherhood of the Serpent".
- **4**-The Phoenicians had a secret society called "The Fish" and it was headquartered in Arwad.
- **5**-The Templar Order used the island as a hide-out. In fact, Arwad sheltered the last Crusaders and the remnants of the Templar knights who fled France following their massacre on the hand of the king of France and the infamous Inquisition.
- **6**-Some claimed that the Templar knights returned to Arwad to retrieve the Holy Grail; the genealogical tree of Jesus and a buried gospel by Mary Magdalene. Ironically, at one time in history, Arwad was Pope Clement V's gift to the Templar knights.
- **7**-The Island of Arwad was the last stronghold of the Crusaders in the Near East.
- **8**-During the French occupation of Syria and Lebanon during the Second World War, the Vichy French Government discovered Anunnaki-Phoenicians tablets buried underground in Arwad. The ancient tablets told the story of a race of super humans who descended on earth and taught the fishermen how to navigate the sea and how to read the maps of the stars.

- **9**-They created a secret society called "Brotherhood of the Fish".

> Later on in history, the "fish" became the secret symbol of early Christians. It was St. Paul who first created the fish symbol as a secret way for early Christians to recognize other Christians in the Levant, Greece and Rome. In ancient times, Arwad was a refuge to many persecuted Phoenicians, Hebrews, as well as Greeks and Romans. It is a perfect spot for a modern time Ernest Hemingway. You will be intrigued by the layout of the island's houses and its fortress.

What is Ay'inbet?

Ay'inbet is an Ana'kh/Phoenician term for:
- **a**-A chosen group;
- **b**-A social class;
- **c**-Favorite subjects;
- **d**-Those who know.

It is derived from the Phoenician Ay'inbet. Ayin means eye, and Bet means house.
- In Hebrew, Ayin is Ayn and bet is beth.
- In Arabic, Ayin is 'ayn, and bet is bayt (Written Arabic), and bet (Spoken Arabic).
- In Ana'kh, it is exactly the same; Ayin is Ain, and Bet is Bet.

The general meaning is:
- **1**-The eye of the house;
- **2**-Main entrance of a home;
- **3**-Protection of one's home.

Baalshalimroot

The upper class of the Anunnaki is ruled by Baalshalimroot. His subjects are called "Shtaroout-Hxall Ain", (Shtatroot Kha-Ayn" meaning the inhabitants of the house of knowledge, or Tthose who see clearly." Their eyes are not similar to humans' eyes, because the Anunnaki do not have a retina. Their physical eyes are used to perceive dimensional objects. While their "inner eye" sees multi-dimensional spheres.

The process is created by the mind. The word Ain was later adopted by the early inhabitants of the Arab Peninsula. Ain in Arabic means eye.

The Badou Rouhal (Nomads) of the Arabs who lived in the Sahara considered the "eye" to be the most important feature of the face.

Those who have practiced As Sihr (Magic) used their eyes as a psychic conduit

Those who have practiced As Sihr (Magic) used their eyes as a psychic conduit. In their magic rituals and séances, they close their eyes and let imageries inhabit their mind.

Once, the spirit called Rouh, Jinn, Afrit enters the body, the eyes open up and the vision is henceforth activated by the spirit. And what they saw next was called Rou'Yah, meaning "visions".

In the secret teachings of Sufism, visions of Al Hallaj, and of the greatest poetess of Sufism, Rabiha' Al Adawi Yah, known also as "Ha Chi katou Al Houbb Al Ilahi" (The mistress of the divine love), and in the banned book Shams Al Maa'Ref Al Kubrah (Book of the Sun of the Great Knowledge), the word eye meant the ultimate knowledge, or wisdom from above. "Above" clearly indicates the heavens. In the pre-Islamic era, heavens meant the spheres where the creators of the universe live. This sphere was shared by good gods and evil gods.

The concept of hell was unknown to the pagan Badou Rouhal

The concept of hell was unknown to the pagan Badou Rouhal. Later on in history, when Islam invaded the Arab world, the eye became the symbol of Allah, the god of the Muslims. In modern times, several secret esoteric societies and cultures adopted the eye as an institutional symbol and caused it to appear on many edifices' pillars, bank notes, money bills (Including the US Dollar), and religious texts. In ancient times, the Anunnaki eye was a very powerful symbol of the favorite regional god.

It appeared on Egyptian, Sumerian, Persian, and Phoenician pillars and tablets. The Phoenicians of the city of Amrit and the Island of Arwad, considered to be direct descendants of the Anunnaki, engraved the Anunnaki eye on altars dedicated to gods' healing powers. Code/Use according to mythology and esoterism: To be written three times on a piece of leather or cloth and hidden in the left pocket.

Esoteric benefits:
- 1-Safe return to home-base;
- 2-Against forced eviction;
- 3-Protection of one's property;
- 4-Peace at home.

Geometrical presentation/Symbol: Circle. In spiritual-mental séances, the circle becomes a triangle.

What is Azaridu?

Azaridu is an Ana'kh name for a sort of liquid medicine. According to the Ulema, a similar liquid was used by the early Phoenicians of Arwad and Ugarit to heal wounds and cast out evil spirits. Later on in history, the Arab alchemists mixed this mysterious liquid with wild plants and Zaafaran to transmute objects and metals into celestial elements. This is of course a myth. However, Ulema As Sadik said: "The Zaafaran was an essential ingredient in all spiritism séances, and the Azaridu was used in fact, but not to transmute metals or objects."

*** *** ***

From the book "Anunnaki Dictionary Thesaurus: Ana'kh Language of the extraterrestrial gods and goddesses"
By M. de Lafayette

A selection of important words from the Anunnaki's language.

Aamala: Anunnaki's registry of future events. It is used as a calendar to show important events that will occur on other planets. According Ulema Rajani, time is not linear. And because space bends on itself, therefore, events don't have a chronology or time-sequences. "Things and events happen on the net of the cosmos. When your mind perceives them, they happen before your eyes. But in fact, they have already happened before your have noticed them. This applies to all future and forthcoming events, because also they have occurred on another cosmos net parallel to the one that have contained separate events. It is a matter of perception, rather than observation or taking notice..." said Ulema Govinda.

Abamarash: Anunnaki's spatial transmission of thoughts on a holographic grid. The thoughts register as codes on an electro-plasmic screen or mirror. The Abamarash codes appear in sequences of numbers, usually a multitude of dots and lines. The dot corresponds to zero. The line corresponds to 1.

Abekira'h-Kitbu: Anunnaki's act/process of recording recently acquired information in the conic books of the central library in Nibiru. Anunnaki's books are made from a plasmic material/substance that resembles aluminum. They have a conic shape, and they rotate on shelves. They are opened or activated either by thought or by pointing at the digital number of each cone.

Aberuchimiti: Laboratories' tubes used in genetic creation.
It is composed of two words:
a-Aberu, which means container; tube.
b-Chimiti, which means a laboratory.
The word Chimiti appeared in Sumerian epics and texts.
See Chimiti.

Abgaru: Balance; equilibrium. Balance does not mean a physical balance, but a position or a situation where and when a person maintains a perfect vision, assimilation and understanding of the limits, dimensions and length of objects surrounding him/her.
In other words, it is sensing and remembering the exact position of objects that can expand within the area where we are standing or walking, even in the dark.Objects are not limited by their physical dimension, and/or the physical place they occupy. "Almost all objects extend and expand outside what it defines their measurement, shape, and size, because all objects have inertia "Energy" rays or vibes that constantly emanate from them, thus occupying an extra physical place. Not to bump into the vibes area is maintaining balance," said Ulema W. Li.

Abra-ah: It is the name of the Fiffth Element in the Anunnaki's matrix. It means transcending time and space. The Ulema coined it "Niktat Alkhou-Lood", and it means verbatim: The point of the beginning of immortality. In other words, that Fifth element is responsible for an extraordinary longevity of mankind on earth, and/or its immortality.
It is very possible said an Ulema that after 2022, humans will learn about the secret of immortality, but will never be able to decode the composition and sequences of the fifth element. It would be a catastrophe for humanity and for the future of planet earth, if humans succeed in decoding the data contained in the fifth element.
Many Ulema are not seriously worried, because with the arrival of the Anunnaki in 2022, the existence of human life and its continuity will be in the hands of the Anunnaki.

Akama-ra: According to the Ulema, the Akama-ra were the first beings who were allowed by the Anunnaki (Enki and Inanna) to date the "Women of Light" who were quarantined on earth by the Anunnaki. Akama-ra were genetically created by the Anunnaki on Ashtari (Nibiru) and were transported to planet earth on Anunnaki's spaceships, called Merkabah.

Anšeduba: The entire hell; the several levels of the underworld. Note: It is defined in terrestrial terminology, and as it is used or understood in the Akkadian/Sumerians clay tablets. "Anšeduba is a metaphoric word, because the concept of hell does not exist in extraterrestrial literature and cosmology. Hell is the product of the human mind, created by frightened and confused early humans, so-called prophets and preachers. And promoted by organized religions for their own profit, and for establishing control over the mind, the soul, and the body of duped followers..." said Ulema Nejad.

He added: "This takes me to questioning the identity and personality of the scribes and writers of the Akkadian/Sumerian clay tablets...where they humans, messengers of the gods, or extraterrestrial Anunnaki who manifested themselves on earth as kings and builders of civilizations?

The Akkadian/Sumerian epics mentioned the descent of gods and goddesses to the underworld...this means one thing to me: The epics were written by humans and inspired by terrestrial beliefs...and rotated around physical beings, people from flesh and blood who were human beings not the Anunnaki, the very advanced extraterrestrial race.

The Anunnaki did not believe in hell, because hell exists only in the mind of weakened and frightened human beings. This also makes me believe that those epics are neither the accumulation of historical facts, nor the accurate and entire story of the Anunnaki's existence on Earth...they are simple poetry."

Anubdada: The seven dimensions or planets visited by the Anunnaki, and described in the Book of Rama-Dosh. It is understood by the adepts and the enlightened ones that the Anunnaki have visited a great number of stars and planets. Their trip to planet Earth was neither the only voyage they have made, nor the mot significant one.

Ulema Sorenztein said: "As a matter of fact, their (Anunnaki) visit to Earth was the less significant one...you have to remember that at the dawn of the existence of Earth, no civilizations were established, art did not exist, poetry was not written yet, science not found yet, in brief, an archaic form of life roamed its surface and metamorphosed beneath...Thus, it is quite safe to assume that Earth at that time in history had little to offer to very advanced galactic civilizations. So why did the Anunnaki land on Earth?

There are of course many theories and assumptions on the subject.

Few archeological remains and facts attest to their arrival, however, not always authoritative and convincing. But what is certain is the fact that the Anunnaki like so many other extraterrestrial civilizations have visited other galactic systems and established some sort of relations and enterprises with other galactic societies.

It was part of their routine activities and the very nature of their existence. So far, a few authors have suggested that the Anunnaki had extensive rapports with aliens from Zeta Reticuli, but this is the tip of the iceberg. There are billions upon billions of stars and planets in the universe, and Zeta Reticuli and Earth are but a dot on the cosmic map.

The reason for mentioning the seven planetary systems or spheres in the Book of Rama-Dosh is not clear, but I think they were mentioned because at the beginning of time as we know it or understand it, the Anunnaki have opened Ba'abs (Stargates) that allowed them to zoom in the universe faster than the speed of light. One of these gigantic Ba'abs was located not far from planet Earth.

Another stargate was located near Mars. Between the two stargates, the Anunnaki created the Plasmic Belt, also known as the Spherical Shield. Two of the seven destinations mentioned in the Book of Rama-Dosh were Mars and Earth."

An-zalubirach: Collecting thoughts and multiple mental images; using mental energy to move or teleport things. This is one of the phases and practices of Tarkiz.

Tarkiz means deep mental or intellectual concentration that produces telekinesis and teleportation phenomena. Anunnaki's young students, apprentices, and novices learn this technique in Anunnaki schools in Ashtari.

Usually, they use their Conduit (Which is located in the brain's cells), and deep concentration on an object hidden behind a screen or a divider made from thin paper. Synchronizing the frequency of their Conduit and an absolute state of introspection, the Anunnaki student attempts to move the hidden object from one place to another without even touching it. In a more advanced stage, the Anunnaki student attempts to alter the properties of the object by lowering or increasing the frequencies and vibrations of the object.

Anzuma: Anunnaki's designated spots or areas of landing on earth. Usually, they are used to map terrains to serve as landing terminals. The first time the Anunnaki visited planet earth, they marked the regions where they have landed their Merkabah (Spaceships). These regions were the ancient Phoenicians cities of Tyre, Sidon Afka, Baalbeck (Baalback), Bijeh, Amchit, Byblos, Batroun and Arwad Island.

Baliba nahr usu na Ram: Translated verbatim: "The water of the river purified my people." Attributed to Sinhar Marduchk in the Book of Rama-Dosh. Baliba means flows of waters. Nahr means river (Same meaning in Hebrew, Phoenician and Arabic). Usu means to clean or purify. Na means my or our. Ram means people (Same meaning in Phoenician, primitive Arabic, early Armenian and ancient Hebrew). The Ana'kh word "Usu" also means to dig. We find similar meaning in the Annals of Sardanapalus: "Nahrtu istu nahr zaba anta ahri nahr babilat kanin sumsa abbi." Translated verbatim: "A river from the upper Zab I dug and its name I called."

Balu-ram-haba: Composed of three words:
a-Balu, which means power; transition; contact.
b-Ram, which means oeople. In this case, entities; other life-forms.
c-Haba, which means beyond; other dimension.

Possibly, from Balu-ram-haba, derived the Hebrew word Olam ha-ba. This Ana'kh term or expression pertains to circumstances in the world beyond, and/or experiences, the departed humans might encounter in the next dimension, following their death. On this subject, the Anunnaki-Ulema have said (Excerpts from their Kira'at, verbatim): Afterlife does not necessarily begin after we die, because death does not exist; it is simply a transitory stage. Within our physical world exist so many other worlds. And far away, and deep in the fabric of the universe, distances are reduced, even eliminated, if we zoom into our Double. Matter and anti-matter are de-fragmented in the parallel dimension.The initiated and enlightened ones can transport themselves to the other world, and visit the far distant corners of the universe through their Double. Those who are noble in their thoughts, intentions and deeds can accomplish this after an Ulema initiation.
- The righteous people will be reunited with their loved ones including their pets in the afterlife.

- This reunion will take place in the ethereal Fourth dimension.
- The reunion is not of a physical nature, but mental. This means, that the mind of the deceased will project and recreate holographic images of people, animals and places.
- All projected holographic images are identical to the original ones, but they are multidimensional.
- Multidimensional means, that people, animals and physical objects are real in essence, in molecules, in DNA, and in origin, but not necessarily in physical properties. In other words, what you see in the afterlife is real to the mind, but not to your physical senses, because in the after life (In all the seven levels/dimensions of life after death), physical objects, including humans' and animals' bodies acquire different substances, molecular compositions, and new forms.
- The physical rewards and punishments are mental, not physical in nature, but they are as real as the physical ones.
- The deceased will suffer through the mind.
- The pain sensations are real, but are produced by the mind, instead of a physical body. So in concept and essence, the Ulema and Hebraic scholars share similar beliefs; the good person will be rewarded, and the bad person will be punished.
- For the Jews, it is physical, while for the Ulema it is mental, but both reward and punishment are identical in their intensity and application.
- The wicked will not be indefinitely excluded from a reunion with loved ones.
- The wicked will remain in a state of loneliness, chaos, confusion and mental anguish for as long it takes to rehabilitate him/her.
- This state of punishment and rehabilitation can last for a very long period of time in an uncomfortable sphere of existence inhabited by images of frightening entities created by the mind as a form of punishment.
- Eventually, all wicked persons will reunite with their loved ones after a long period of purification and severe punishment.

Here are the views of the Anunnaki-Ulema:

- Soul is a metaphysical concept created by Man.
- Soul is a religious idea created by humans to explain and/or to believe in what they don't understand.
- It is more accurate to use the word Mind instead.
- The mind thinks and understands. The soul does not, perhaps it feels, if it is to be considered as a vital force and source of feelings in your physical body.
- In the afterlife, such source of feelings is non-existent, and in the dimensions of the after world, such source is useless.

Barak yom-ur: Verbatim translated: "Blessing or decree on the day of the city." The general meaning is the blessing received during the memorial day of the establishment of a city. It is composed from three words:
a-Barak, which means blessing.
b-Yom, which means day; memorial day or moment.
c-Ur, which means city.
From the Ana'kh Yom, derived the Hebrew and Arabic word Yom, which means day. From the Ana'kh Barak, derived the Hebrew and Arabic words Barak, Barakat, and Barakaat, which mean blessing.

Barak-malku: Blessing of the ruler or the king; long live the king. Composed of two words:
a-Barak, which means blessing.
b-Malku, which means king.
Barak is Barak and Barakat in ancient Hebrew, and Barak and Barakat in Arabic. Malku is Malku, (Plural: Malki) in Assyrian. In Aramaic, it is Malak (King).
In Hebrew, it is Malek (King). In Arabic, it is Malak (King). From the Ana'kh Malku derived the Aramaic Malkut or Malakut (Kingdon; paradise), the ancient Hebrew Malkuth (Kingdom), and the Arabic Malakoot (Paradise; kigdom of God). Not to be confused with the Semitic words Mala'k or Malak wich mean angel in Hebrew and Arabic.
Sargina said: "Sar sa ultu yom biluti-su malku gabra-su la ispu." Translated verbatim: "King who from the day of his power, a

prince his rival has not been." From the Annals of Tiglath Pileser: "Malki nikrut Assur abil." Translated berbatim: "Monarchs enemies of Assur (Ashur) I seized."

Chavad-nitrin: Ana'kh/Ulemite/Phoenician.
An ancient Phoenician embalming process, learned from Byblos and Arwad Anunnaki remnants, using Mah'rit, a secret substance considered to be humanity's first formula for steroid, and barks from the cedars of Phoenicia (Lebanon, today.)
It was frequently used by athletes in Ugarit, Amrit, and Arwad. From Chavad-nitrin, derived the Greek word Natron or Natrin, a substance used in the embalming process.
Because of the physical strength and longevity the Phoenician's Nitrin produced, the Egyptians equated the words Neter, Netjer and Netjet with the immortality of the gods. According to Egyptian tradition, the first human being to be mummified was god Osiris, whose body was floated down the Nile in a wooden casket and washed ashore at Byblos in Phoenicia. For this reason, Byblos was a sacred region to Osiris, to his cult, and to the Egyptians. This explains the reason why the Egyptians have called Byblos the land of the god Osiris, and Ta Netjer.

Chimiti: Ana'kh/Ulemite/Sumerian/Akkadian. Name of the Anunnaki's laboratory, where humans were created genetically. The Anunnaki's chief geneticist Enqi, and the chief medical observer Ninti used genetic manipulations and in-vitro fertilization of a biological entity in a controlled, non-living environment as a laboratory vessel. Sometimes, glass tubes were used. This process occurred in an Anunnaki's laboratory called Chimiti. The meaning of Chimti is the house where the wind of life is breathing in.

Dab'Laa: Term for the Anunnaki's branching out and changing individuality in multiple universes. This is a very complex topic, and a phenomenon difficult to comprehend. To understand the concept, the closest metaphor in human terms would be – for instance - if you wish you could do something differently, if you could change the past, perhaps changing a previous decision you have regretted, by going back in time to a point before you have made that decision. Through the Dab'Laa, an Anunnaki can split himself/herself in two, or more if necessary, and move on to a universe that is very much like the one he/she lived in.However, if an Anunnaki wishes to branch out and move on, he/she must study the matter very carefully and make the right selection. The branching, or splitting results in exact copies of the person of the Anunnaki, both physically and mentally.

At the moment of separation, each separate individual copy of an Anunnaki grows, mentally, in a different direction, follows his or her own free will and decisions, and eventually the two are not exactly alike.

The old copy stays where he/she is, and follows his/her old patterns as he/she wishes. The new one might land one minute, or a month, or a year, before the decision he/she wants to change or avoid.

Ulema Ghafari Azad said: "Let's take this scenario for instance: Some 30,000 years ago in his life-span; an Anunnaki male was living a nice life with his wife and family. But he felt that he did not accomplish much, and suddenly he wanted to be more active in the development of the universe; a change caused by witnessing a horrendous event such as a certain group of beings in his galaxy destroying an entire civilization, and killing millions of the inhabitants, in order to take over their planet for various purposes.

It happened while an Anunnaki was on a trip, and he actually saw the destruction and actions of war while he was traveling. It was quite traumatic, and he thought, at that moment, that he must be active in preventing such events from occurring again, ever. So, he went back in time to be in a particular spot, where he could prevent these fateful events from happening again. There, in that new dimension, the Anunnaki leaves his/her former self (A copy of himself) as a guardian and a protector. The other copy (Perhaps one of the original ones) is still on Nibiru." This phenomenon is called Dab'Laa.

Dadmim "Admi", "Adamai", "Adami": A human creature; people on earth. From Dadmin and Admi, derived the Assyrian Dadmi, which means mankind and people.

Da-Irat: Anakh/Ulemite. Name for Anunnaki's "Circle Technique"

Da-Irat is known to the enlightened ones and Ulema' adepts as the "Circle Technique" (Da-Ira-Maaref), which means the circle of knowledge. This technique eliminates stress, through one's self-energy. In other words, it is an Ulema technique used to

energize one's mind and body, and to eliminate worries that are preventing an individual from functionining properly everywhere, including the office, home, social gatherings, etc. In the West, zillions of techniques to reduce stress and counter bad vibes were proposed. And many of those techniques work very well.
The following will explain the Ulema's techniques that were in practice for thousands of years in the Near and Middle East. No physical exercise is required. It is purely mental, although some of the steps to follow might look esoteric or spiritual in nature. These techniques were developed by the early Ulema and members of the "Fish Circle", a brotherhood of the ancient island of Arwad, where allegedly, remnants of the An.na.Ki (Anunnaki) lived, and developed the Mah-Rit in sophisticated genetic labs.

The Da-Irat Technique (The Circle Technique) as used by the Fish Circle Brotherhood.
Terminology:
1-In Ulemite, it is called Da-Irat (Circle; sphere).
2-In Ana'kh, it is called Arac-ta.
3-In Phoenician, it is called Teth-Ra. Teth is circle or good thing. Ra is creative energy or first source of life.
Application and Use:
Excerpts from the Anunnaki-Ulema Kira'ats.
Note: Translated verbatim (As Is) from the original text and readings in Ulemite and Phoenician:

- 1-You have to create a space, or find a new space where your mind can manifest itself.
- 2-You live through your mind, and not through your soul.
- 3-Therefore, no spiritual exercise is required.
- 4-A mental exercise is a prerequisite.
- 5-You can create a space for your mind by putting aside for a short moment all your worries, thoughts and other activities.
- 6-You enter your private room, and you sit there for five seconds, doing nothing, and trying to think about nothing.
- 7-Find a comfortable spot.
- 8-Untie your belt and take off your shoes.

- 9-Bring your arms close to your hip. This position is called Kaph.
- 10-If you managed to stop thinking about anything and everything, this would be great. If not, do this:
- 11-Speak to your mind vocally. Do not be embarrassed. Nobody is going to think you are crazy, because you are alone in your room.
- 2-Tell your mind that you want to see right before your eyes a large white door on your right, and a small blue window on your left. This exercise is called Qoph. It is a Phoenician word, and it means the eye of the needle. In Ulemite, it is called Qafra. In Ana'kh, it is called Kaf-ra-du.
- 13-Tell your mind you are knocking at the door, and you are getting ready to walk in.
- 14-Knock at the door. Yes, you can raise your hand and knock on the door. Do not hesitate. You will go through. It is guaranteed.
- 15-Before you enter, look to see if the window is closed.
- 16-If the window is closed, that's fine, open the door and get inside.
- 17-If the window is open, close it. Make sure it is close.
- 18-Now, proceed and get in...
- 19-As soon as you step in, shut the door behind you.
- 20-Your mind is free now.
- 21-Close your eyes. Do not open your eyes, until I will tell you when.
- 22-Your mind is no longer distracted by other things.
- 23-You are now in a state of serenity.
- 24-Your mind is getting settled.
- 25-Your mind is finally finding its place. This stage is called Taw in Phoenician, and it means mark. You are marking now your state of mind.
- 26-This is the place where you are going to dump your worries.
- 27-You start to feel very quite, calm and relaxed. And that is good.
- 28-Keep your eyes closed.
- 29-Now, your mind is ready to listen to your command and wishes.

- 30-Tell yourself you are entering inside your body.
- 31-Tell yourself you want to enter inside your head. This stage is called Resh in Phoenician, and it means head. In Arabic, it is called Ras. In Ulemite is called Rasha. In Ana'kh, it is called Rashat.
- 32-Bring yourself very close to your forehead.
- 33-Direct yourself toward your eyebrows.
- 34-You are seeing now a calm stream of water.
- This stage is called Mem in Phoenician, and it means water. In Ulemite, it is called Ma'. In Arabic, it is called Ma' or Maiy.
- 35-The water is running very smoothly.
- 36-Flow with the stream.
- 37-Become one with the stream.
- 38-The stream is showing you now beautiful sceneries.
- 39-The stream is branching out very gently and is diving itself into small canals. And that is good.
- 40-Can you count how many canals are you seeing?
- 41-Follow the canals and count how many are they.
- 42-Can you try to join them together?
- 43-Try again.
- 44-Let your mind help you.
- 45-Ask your mind to grab all these canals.
- 46-You see, you are doing it. This stage is called Heth in Phoenician, and it means fence.
- 47-You are feeling so good.
- 48-Now, breathe slowly and gently.
- 49-Continue to breath.
- 50-Now tell yourself you want to continue going inside yourself.
- 51-Can you manage to bring the river inside your body?
- 52-Try.
- 53-The river is coming closer to your chest.
- 54-You start to feel it.
- 55-Yes, the canal is entering your chest. And you are feeling so good, fresh, energized.
- 56-Tell the canal to clean everything in your body.
- 57-Everything...everything...

- 58-You are feeling now the fresh and cool water everywhere in your body.
- 59-The beautiful fresh water is cleaning all the mess and dirt inside your body.
- 60-Command the water to flush everything out.
- 61-Tell your mind to close your body.
- 62-Nothing now can enter your body. It is sealed. It is clean and sparkling, and you feel it.
- 63-Now, you want to visit your knees.
- 64-Go there. Turn around your knees.
- 65-Turn one more time.
- 66-You are going to see now white spirals of light surrounding your knees.
- 67-Stay there. Let the light turn and turn and turn around your knees.
- 68-Now the light is going down…down toward your feet, and that is good.
- 69-Your feet feel good.
- 70-Your feet are floating now.
- 71-Let them float.
- 72-You start to feel as if you are sliding gently…
- 73-You are, but you are floating. And that is so good.
- 74-Tell yourself your heart is strong and healthy.
- 75-Tell your lungs they are clean and healthy.
- 76-Tell your body how wonderful and strong it is.
- 77-Tell yourself how wonderful and strong you are.
- 78-Thank your mind for this wonderful journey.
- 79-Tell your body that your mind is standing by your body and is going to take care of it.
- 80-Breathe slowly and deeply three times.
- 81-Tell yourself you have done a good job and you are going to open your eyes now.
- 82-Open your eyes.
- 83-Stretch your arms gently.
- 84-Stay put for 5 seconds.
- 85-Stand up.
- 86-Take a nice hot shower.

Ulema Win Li said:

- Repeat this exercise twice a week.
- The second time you do this, you are going to feel much better, and the exercise will look more pleasant.
- After the second exercise, you are going to notice a great improvement in your mental and physical health.
- Take note of your progress, and compare notes.

Dayyakura: A set of laws governing extraterrestrial relations between various alien races. In other words, Dayyakura is an extraterrestrial/galactic cosmic law. From Dayyakura, derived the Ana'kh word Dayanna, which means a judge. From Dayanna, derived the Hebrew word Dayyan, which means a Talmudic judge, as well as the Arabic word Dayyan (Kadi), which means a ruler, a magistrate, a judge.
From the Ana'kh words previously mentioned, derived several Semitic and Middle/Near Eastern words, such as the Arabic word Din, which means religion or faith, the Sumerian word Deena which means religion, the Aramaic Dino, which means religion and law.

Dhikuru: Anunnaki's secret names or sounds that produce magical effects. Similar to a certain degree to the Islamic 99 secret names of Allah, and to the Kaballa's secret Divine names used to create a Golem.

Eido-Rah: Term for the non-physical substance of a human being's body. In other words, the mental or astral projection of the body leaving earth. Eido-Rah manifests to human beings, and particularly to the parents of the deceased one during a period of less than 40 days, following the death of a relative. From Eido-Rah, derived the Greek word Eidolon (A phantom). According to the "Book of Rama-Dosh": "After we die, the primordial source of energy in our body leaves our body.
This energy is a substance made out of Fik'r closely connected and attached to a copy of ourselves preserved in the Fourth dimension, which is not very far away from us, and from earth. As soon as this energy leaves the physical body, the mind of the deceased becomes confused instantly.
The mind does not realize that the body is dead. At this particular stage, the mind is unable to realize right away that it has entered a new dimension. Although this new dimension is

identical to the one we live in and what we call earth, it is also very different because time, space and distance no longer exist. Everything becomes meta-linear.

Because the mind is confused, it tries to return to earth. The first places, the mind (Or the new form-substance of the deceased one) searches for, and/or tries to return to, are those He/she/It knew best, and familiar with, such as home, office, recreation center, church, mosque, synagogue, temple, etc...but the most sought place is usually home.

So, the deceased one returns home for a very short period. This does not happen all the time. Only when the deceased is totally confused and disoriented. First, the deceased tries to contact relatives and close parents.

When the deceased begins to realize that parents and relatives are not responding, the deceased tries again to send messages telepathically.

Some messages if intensified can take on ectoplasmic forms, or appear as a shadow usually on smooth substances such as mirror and glass.

Some deceased people will keep on trying to contact their beloved ones left behind for a period of 39 days and 11 hours. After this time, the deceased dissipates, and no further attempts to establish contact are made."In another passage of the Book of Rama-dosh, we read (Verbatim): "Although, it is impossible to reach the deceased one as soon as he/she leaves the body, and/or during the 39 days and 11 hours period following his death, sometimes, if we are lucky, and/or were extremely attached to the person we lost, a short contact with him or with her is still possible if we pay attention to unusual things happening around us...those unusual things are difficult to notice, unless we pay a great attention....they happen only once, sometimes twice, but this is very rare..." The book provides techniques and methods pertaining to all forms and means of such contact

Emim: Ana'kh/Sumerian/Hebrew. Name given to the children of Anak. The Bible referred to them as the offspring of the giants and the women of earth. They are the corrupted offspring of the Anakim.

Here are some excerpts from the Bible:

- Jos: 11:21: "And at that time came Joshua, and cut off the Anakim from the mountains, from Hebron, from Debir,

from Anab, and from all the mountains of Judah, and from all the mountains of Israel: Joshua destroyed them utterly with their cities."
- Jos: 11:22: "There was none of the Anakim left in the land of the children of Israel: only in Gaza, in Gath, and in Ashdod, there remained..."
Jos: 14:12: "Now therefore give me this mountain, whereof the Lord spoke in that day; for thou heardest in that day how the Anakim were there, and that the cities were great and fenced: if so be the Lord will be with me, then I shall be able to drive them out, as the Lord said."
- Jos: 14:15: "And the name of Hebron before was Kirjatharba; which Arba was a great man among the Anakim. And the land had rest from war."

Ezakarerdi, "E-zakar-erdi "Azakar.Ki": Ana'kh/Ulemite. Term for the "Inhabitants of Earth" as named by the Anunnaki, and mentioned in the Ulemite language in the "Book of Rama-Dosh." Per contra, extraterrestrials are called Ezakarfalki. "Inhabitants of Heaven or Sky". The term or phrase "Inhabitants of Earth" refers only to humans, because animals and sea creatures are called Ezbahaiim-erdi. Ezakarerdi is composed of three words:
1-E (Pronounced Eeh or Ea) means first.
2-Zakar: This is the Akkadian/Sumerian name given to Adam by Enki. The same word is still in use today in Arabic, and it means male. In Arabic, the female is called: Ountha (Oonsa).
The word "Zakar" means:
a-A male, and sometime a stud.
b-To remember.
In Hebrew, "Zakar" also means:
a-To remember (Qal in Hebrew).
b-Be thought of (Niphal in Hebrew).
c-Make remembrance (Hiphil in Hebrew).

There is a very colorful linguistic jurisprudence in the Arabic literature that explains the hidden meaning of the word "Zakar"; Arabs in general believe that man (Male) remembers things, while women generally tend to forget almost everything, thus was born the Arabic name for a woman "Outha or Oonsa", which means literally "To forget!" Outha (Oonsa) either derives from or coincides with the words "Natha", "Nasa", "Al Natha", "Nis-Yan", which all mean the very same thing: Forgetting; to forget, or not to remember. On a theological level, Islamic scholars explain that the faculty of remembering is a sacred duty for the Muslim, because it geared him toward remembering that Allah (God) is the creator. Coincidently or not, Zakar in Ana'kh (Anunnaki language) and ancient Babylonian-Sumerian means also to remember. Could it be a hint or an indication for Adam's duty of remembering Enki, his creator?

3-Erdi means planet Earth. Erdi was transformed by scribes into Ki in the Akkadian, Sumerian and Babylonian epics.
From Erd, derived:
a-The Sumerian Ersetu and Erdsetu,
b-The Arabic Ard,
c-The Hebrew Eretz.

All sharing the same meaning: Earth; land.Thus the word Ezakarerdi means verbatim: The first man (Or Created one) of Earth or the first man on Earth, or simply, the Earth-Man. In other word, the terrestrial human.

Ezakarfalki "E-zakar-falki": Term for extraterrestrials as mentioned in the "Book of Rama-Dosh." Per contra, inhabitants of planet Earth are called Ezakarerdi or Ezakar.Ki.

Ezbahaiim-erdi: Term for all the animals, and sea creatures living, and/or created on planet Earth. It is composed of three words:
a-Ez means first creatures of a second level. (In comparison to humans.)
b-Bahaiim means animals. The same word exists in Arabic, and means the very same thing (Animals).
c-Erdi means planet Earth.

Ezrai-il: Ana'kh/Ulemite: Name of super-beings, who can transcend space and time, and appear to human as angels, in terrestrial term. The Ana'kh literature refers to them as ethereal manifestation of the matter. But, our religions and Holy Scriptures depict them as the fallen angels. Ezrai-il or Ezrail is composed of two words:

a-Ezra, which means message or manifestation.
b-Il, which means divine; god; creator(s).

F "ف": One of the esoteric letters in the Anunnaki, Ulemite and Arabic alphabets. Typical with ancient Semitic names, there are none that begin with an 'F'.
The Ulema called "F" the forbidden letter, or more precisely, the letter that was never allowed to be included in the Phoenician Aramaic, and Hebrew alphabets. "Thus all secret sounds and meanings associated with F would not be pronounced or heard, or known to the un-enlightened ones..." said Ulema Hanafi. "There are 12 secret words starting with the letter F that are hidden in the Torah, and the Book of Rama-Dosh...", explained Ulema Sadiq Al Qaqsi. And accordingly, each word produces a powerful sound capable of changing the fabric of time. The letter F was substituted by Ph, pronounced Pveh in several Semitic languages, except Arabic; the proto-Semitic "P" became the Arabic "F".

Fari-narif "Fari-Hanif": Or simply Ra-Nif. A term for categorizing different forms of spirits, or non-physical entities. The Anunnaki referred to many different forms, shapes and "rating" of entities known to the human race as "Spirits" and "Souls". From "Ra" and "Ra-Nif", derived the Semitic words Rouh, Rafesh, Nefes, Nefs, Roach, and Ruach, meaning the soul in Arabic, Aramaic and Hebrew.
Souls and spirits in Anunnaki-Ulema literature: "The soul or spirit is a concept created by man," said Ulema Bakri. "Man was created, lives and continues to live through his Mind...not through his soul. Angels do exist, but humans don't understand a thing about them, because religions taught Man that angels are spirits. The truth is they are neither spirits, nor the messengers of God, but a projection of a higher level of goodness and intelligence...and those who (after their physical death on earth) enter a higher dimension, would meet the angels in the Fourth dimension as meta-plasmic presence emanating beauty and goodness, but they are not divine spirits or pure souls..." added Ulema Bakri.

Ghen-ardi-vardeh "Gen-adi-warkah": Aagerdi-deh for short. The act or process of talking to others without using words, and in a total silence. Composed from three words:

a-Gen, which means people; others.
b-Ardi, which means earthly; land; location. Ertz in Hebrew, and Ard in Arabic.
c-Vardeh, which means rose; flower; aroma; chalice; quest. Vardeh in Hebrew and Arabic, and it means a rose in both languages.Warkah is a substitute for Vardeh, and it means a paper; a page.
Ulema Seif Eddine Chawkat told a great story about Gen-ardi-vardeh, and briefly explained how it works in the Book of Rama-Dosh. He said (Verbatim): During World War One, my father worked as a military superintendent for the Turkish army. In some prisoners of war camps, some medical visits and check-ups were scheduled once a month.
It did not happen in all the concentration camps and centers of detention, but it did happen at one particular place, and my father worked there. Some British officers (Prisoners of war) were treated properly, while others were not so lucky. Malaria, dysentery, and other health problems among prisoners were frequent.
The Turkish army ran out of medicine and quinine pills, and the prisoners' health condition began to deteriorate. In brief, not all prisoners received medical treatment. One of those unfortunate British officers was Major V. H. He was in serious trouble. And because he fell so ill, he could not talk anymore, and succumbed to a threatening fever.
Yet, no medical attention was given to him, until, and probably by pure coincidence or luck, a military doctor entered the tent and saw him there agonizing in his bed. The doctor noticed his serious condition and approached him. Unfortunately, nothing could be done; medicine and pills were no longer available, and the only thing one could have done in similar situations was to wipe out the sweat of the sick prisoners. But something very unusual happened there.
The doctor briefly examined the major who was agonizing, but could not utter not one word. He placed his hand on his forehead and throat, and all of a sudden, as my father recalls, "the whole situation changed immediately.
The doctor ordered one of his adjutants to fetch a certain box; we did not know what it was. When the adjutant returned carrying the box, we saw what was in it; bandages, medicine, pills, everything you needed...syringes...etc. The major was lucky; the doctor took care of him, he gave him a few pills, told

the adjutant to watch over him, and asked him to bring the major new clean pillows and blankets. We were stunned. All of a sudden, the doctor and the major became friends. Two years later, when the war ended, and by a pure coincidence, my father met the military doctor in Budapest, and they start to talk about the war and so many other things.

One of those things was Major V.H. story. My father asked the doctor why he cared so much for the major, and not the others. And the doctor replied that the major was one of the "Brothers". In other words, a novice-Ulema just like himself. How did he find out? Smiling, the doctor told my father: "I touched his forehead and his throat...and by touching his throat I could read his silent message to me.

He told me that he was an Ulema." In other words, the British major used the technique of Aagerdi-deh to talk to the doctor without opening his mouth, and of course to let him know that he was an Ulema.

The British major hoped that the Turkish doctor might be an Ulema himself, and if so, he would be able to get his "silent message." And HE WAS! Both were Ulema, and this is why the Turkish physician cared so much for the British prisoner."Of course nobody believed this", said my father.

In fact, every time my father told this story, people laughed at him."Amazingly, ninety years later, NASA began to explore Aagerdi-deh. They call it now the "Subvocal Speech." Most recently, NASA issued a press release to that effect; it is self-explanatory. Herewith an excerpt from the release, and update on NASA's most fascinating project. "NASA develops system to computerize silent, "Subvocal Speech": NASA scientists have begun to computerize human, silent reading using nerve signals in the throat that control speech.

In preliminary experiments, NASA scientists found that small, button-sized sensors, stuck under the chin and on either side of the "Adam's apple," could gather nerve signals, and send them to a processor and then to a computer program that translates them into words. Eventually, such "subvocal speech" systems could be used in spacesuits, in noisy places like airport towers to capture air-traffic controller commands, or even in traditional voice-recognition programs to increase accuracy, according to NASA scientists. "What is analyzed is silent, or subauditory, speech, such as when a person silently reads or talks to himself," said Jorgensen, a scientist whose team is developing silent, subvocal speech recognition at NASA's Ames Research Center, Moffett Field, Calif. "Biological signals arise when reading or speaking to oneself with or without actual lip or facial movement," Jorgensen explained.

A person using the subvocal system thinks of phrases and talks to himself so quietly, it cannot be heard, but the tongue and vocal chords do receive speech signals from the brain," Jorgensen said.In their first experiment, scientists "trained" special software to recognize six words and 10 digits that the researchers repeated subvocally.

Initial word recognition results were an average of 92 percent accurate. The first sub-vocal words the system "learned" were "stop," "go," "left," "right," "alpha" and "omega," and the digits "zero" through "nine."

Silently speaking these words, scientists conducted simple searches on the Internet by using a number chart representing the alphabet to control a Web browser program.We took the alphabet and put it into a matrix -- like a calendar. We numbered the columns and rows, and we could identify each letter with a pair of single-digit numbers," Jorgensen said. "So we silently spelled out 'NASA' and then submitted it to a well-known Web search engine.

We electronically numbered the Web pages that came up as search results. We used the numbers again to choose Web pages to examine. This proved we could browse the Web without touching a keyboard," Jorgensen explained. Scientists are testing new, "noncontact" sensors that can read muscle signals even through a layer of clothing.

A second demonstration will be to control a mechanical device using a simple set of commands, according to Jorgensen. His team is planning tests with a simulated Mars rover. We can have the model rover go left or right using silently 'spoken' words," Jorgensen said. People in noisy conditions could use the system when privacy is needed, such as during telephone conversations on buses or trains, according to scientists. "An expanded muscle-control system could help injured astronauts control machines. If an astronaut is suffering from muscle weakness due to a long stint in microgravity, the astronaut could send signals to software that would assist with landings on Mars or the Earth, for example," Jorgensen explained. "A logical spin-off would be that handicapped persons could use this system for a lot of things. To learn more about what is in the patterns of the nerve signals that control vocal chords, muscles and tongue position, Ames scientists are studying the complex nerve-signal patterns. We use an amplifier to strengthen the electrical nerve signals. These are processed to remove noise, and then we process them to see useful parts of the signals to show one word from another," Jorgensen said.

After the signals are amplified, computer software "reads" the signals to recognize each word and sound. The keys to this system are the sensors, the signal processing and the pattern recognition, and that's where the scientific meat of what we're doing resides," Jorgensen explained.

We will continue to expand the vocabulary with sets of English sounds, usable by a full speech-recognition computer program. The Computing, Information and Communications Technology Program, part of NASA's Office of Exploration Systems, funds the subvocal word-recognition research. "There is a patent pending for the new technology. (Source: NASA).

Gens "Jenesh": Gender. Similar words appeared in Semitic languages.
To name a few:
a-Gens in Arabic;
b-Gensa in Assyrian;
c-Gensu in Akkadian.

Gensi-uzuru: Apparition of deceased pets. The Ulema are very fond of animals. Extensive passages in the Book of Rama-Dosh speak about the important role animals play in the life of humans, especially at emotional and therapeutic levels. The Ulema believe that pets understand very well their human-friends (Instead of using the word "owners"). And also, pets communicate with those who show them love and affection. This loving relationship between pets and their human-friends does not end when pets die. Although the Anunnaki-Ulema do not believe in any possibility of contacting deceased people or animals, they have explained to us that contacting our departed ones is possible for a very short time, and only during the 40 days period following their death.In other words, we can contact our deceased parents and dear ones, or more accurately enter in contact with them if:
a-They contact us short after their death;
b-They must initiate the contact;

c-This should happen during a 40 days period following their departure;
d-Their contact (Physical or non-physical) must be noticed by us. This means that we should and must pay an extra attention to "something" quite irregular or unusual happening around us. Because our departed pets will try to send us messages, and in many instances, they do.
e-We must expect their messages, and strongly believe in those messages.

The Ulema said that humans cannot contact their dead pets. But pets can contact us via different ways, that we can sense if we have developed a strong with them. Pets know who love them and those who don't, because pets feel, understand, sense and see our aura. All our feelings and thoughts are imprinted in our aura, and the aura is easily visible to pets, particularly, cats, dogs, parrots, lionesses, pigs, and horses.
This belief is shared by authors, people of science and therapists in the West, despite major difference between Westerners and Ulema in defining the nature and limits of pets-humans after death contact.
For instance, in the United States, pets lovers and several groups of therapists and psychics think that "a pet can reappear as a ghost, and a ghost could be luminous or even appear as it did in life. You don't necessarily know when you see an animal if it's a ghost or not, said Warren, a researcher in the field.
"It's much easier to identify a loved one who's passed and come back.""Don't forget them because they're gone," said Jungles, who owns three cats. "Keep their toys and blankets around. They (ghosts) will go where they're happiest."
Warren agrees. "Recreate an environment conducive to the pet's life," he said. "Use your imagination and treat it like it's alive. In other words, you should create or re-create conditions ideal for their re-appearance, even though, for a very short moment.

Ghoolim: A non-physical duplicate of the physical body as projected in the air. More precisely, a holographic picture of the dead body, short after death. From Ghoolim, derived the proto-Arabic word Ghool, which literally means demoniac spirits-beings haunting those who by night visit the cemeteries.

Giabiru: Death; a dead person lost in a parallel dimension. From Giabiru, derived the Assyrian noun Giabi, which means *a reaper*. It did appear in the Akkadian and Sumerian clay tablets."Matani sabzute va malki aibi-su kima giabi uhazizu."-From the Annals of Sardanapalus. Translated verbatim: "Countries turbulent and kings his enemies like a reaper he cut off."

Gibbori: A group of Anunnaki geneticists and people of science who develop DNA sequences, and alter the genes of hybrids. From Gibbori, derived the Arabic word Gabbar which means giants, and the plural Gababira (Giants). In Pre-Islamic era, the word Gababira meant huge entities who came from a non-physical world, and maliciously interfered in humans' affairs. In Hebrew, it is Gibborim (גבר גבור), which is the plural of Geber, which means mighty man. It appears more than 150 times in the Jewish Tanakh.

Gibishi: Power. From Gibishi, derived the Assyrian word Gibis, which means might; power; strength. It did appear in the Akkadian and Sumerian clay tablets. "Mili kassa mee rabuti kima gibis tihamti usalmi."-Nebuchadnezzar. Translated verbatim: "A collection of great water like the might of the sea I caused add it."
"In gibis libbi-ya u suskin galli-ya er asibi."-From the Annals of Sardanapalus. Translated verbatim: "In the strength of my heart, and steadfastness of my servants, I besieged the city.""Ana gibis ummani-su mahdi ittagil."-From the Obelisk of Nimrud. Translated verbatim: "To the powers of his great army he trusted." "Ina gibis emuqi sa Asur bil-ya."-From the Annals of Tihlath Pileser. Translated verbatim: "In the boubdless might of Assur my lord."

Gibsut-sar: A leading group in charge of military operations. Usually, the group consists of five persons, men and women selected from Ma'had, an Ana'kh word meaning an academy. Similar Assyrian word Gibsut-sun appeared in Iraq's ancient clay tablets. In Assyrian, Gibsut-sun means "all of them", referring to groups and gatherings."Kitru rabu iktera itti-su gibsut-sun uruh Akkadi izbatunu."-From a Sennacherib's cylinder. Translated verbatim: "A great gathering was gathered, and with him all of them the road of Akkad took."

Gilgoolim: The non-physical state of a deceased person, at the end of the 40 days period. At that time, the deceased person must decide whether to stay in the lower level of the Fourth dimension, or head toward a higher level of knowledge, following an extensive orientation program/guidance.
From Gilgoolim, derived the Kabalistic/Hebrew word Gilgoolem referring to the cycle of rebirths, meaning the revolution of souls; the whirling of the soul after death, which finds-no rest until it reaches its final destination. But in the Jewish literature and teachings, the final destination is the land of Palestine, the "Promised land".

Ginidu: An enemy.From Ginidu, derived the Assyrian/Akkadian word Gini, which means enemies. It did appear in the Akkadian and Assyrian tablets. "Usanqitu gini Asur."-From the inscriptions of Tiglath Pileser.
Translated verbatim: "He hath subdued the enemies of Assur "Ashur".

Girzutil: Damaged. It could apply to a person or to an object. Most likely, it refers to a damaged region or to a destroyed piece of land. From Girzutil, derived the Assyrian/Akkadian word Girzuti, which means damaged or ravaged. It did appear in the Assyrian tablets. "Eli agari-sun girzuti saharrata adbuk."-Sennacherib. Translated verbatim: "Upon their ravaged fields blackness I left."

Goirim-daru: The vibes produced by one's double, according to the Book of Rama-Dosh; a sort of bio-plasmic rays that project the non-physical properties of an object or a thought.

Golibu "Golibri": Term for the passage or transition from a physical existence to a mental or non-physical sphere, usually associated with the first dimension of the Anunnaki's Shama, meaning sky; outer space; a parallel dimension.

Golim: A prototype of a created presence or entity, usually associated with the mixture of a terrestrial element and the thought of a Golimu who creates a non-human creature. From

Golim, derived the Kabbalistic/Hebrew word Golem. See Golem and Golimu.

Gomari "Gumaridu": A term referring to an Anunnaki Ulema technique capable of manipulating time. It is also called the "Net Technique". Ulema Rabbi Mordechai said: "Human beings treat time as if it were linear.
Day follows day, year follows year, and task follows task. The Anunnaki Ulema, however, have long ago learned how to treat time nonlinearly, and thus be able to accomplish more in their lives." Note: Ulema Rabbi Mordechai is talking to his student Germain Lumiere, who visited with him in Budapest, where he resided for some years.
(Excerpts from the books "On the Road to Ultimate Knowledge", and "The Book of Rama-Dosh, both co-authored by Ilil Arbel and M. de Lafayette.) "It would be beneficial if you could manipulate time in such a way as to be faster than normal people, and this is what we are going to do in the forthcoming exercise.

I. The Exercise:
For the purpose of this exercise, one must have complete privacy, and in addition, one's consciousness changes under the influence of the exercise to such an extent that a mother, for instance, would not hear her children if they need her.
So the exercise cannot be done while young children are at home. Also, if you are taking care of an ill or elderly relative, you should not pursue it either.
If part of the tasks you wish to accomplish are to be done at work, again, you cannot accomplish that because almost all jobs involve the presence of other people. Therefore, for the purpose of this exercise, we will choose a simple frame and an acceptable set of tasks. Let's choose a Saturday, and you have to accomplish a few tasks. All of them must be done on Saturday, because on Sunday you are expecting to be busy with other things. You have, in short, seven hours. Let's assume you have chosen these tasks:
- You have to drive your spouse to the airport.
- You have committed yourself to your boss, promising that you will write a report of a hundred pages or so for Monday.
- You want to shop for food for the week.

This is quite a lot to do in the seven hours that we will assume are available to you during that day. The trip to the airport would take about an hour. The shopping will take about an hour and a half. As for the report, it looks like it should take at least ten hours. So obviously some of the things you wanted to do will not get done. But the Anunnaki-Ulema say that all these things can be done if you learn to break the mold of the linear time, and they have a technique one can learn to do so.

II. The Equipment:
For this technique, you will need a few props:
- A round net. It can be anything – a fishing net, a crochet tablecloth, anything made of thread or yarn with perforations. It should be around four feet in diameter.
- Paper
- Pencil
- Scissors

III. The Technique:
- **1-**Since one of the tasks involves taking your spouse to the airport, work on the preliminary preparations behind a closed door.
- **2-**Look intently at the net, and memorize the way it looks, so that you can easily visualize it.
- **3-**Close your eyes and visualize the net.
- **4-**In your mind, draw a large circle on the net.
- **5-**In your mind, let the net float in the air, making sure it is not flat and horizontal, but moving, bending, waving, and being in a vertical position most of the time.
- **6-**In your mind, concentrate on the tasks you wish to accomplish.
- **7-**In your mind, represent each task as a hole that you mentally perforate in the net. Since you have three tasks, you visualize three holes.
- **8-**Open your eyes, take the physical net, and toss it lightly on a chair or a couch nearby. Do not make it flat and horizontal, just let it land on the piece of furniture like a casual throw.
- **9-**Close your eyes again, and visualize the holes in the mental net. Look at the holes you made, visualizing their

shape, their edges, and their exact position on the mental net.
- **10-**In your mind, throw the mental net on the physical net.
- **11-**Take the paper and pencil, and draw three circles that would match, by their shape and size, the mental holes you have visualized.
- **12-**Cut the circles with the scissors.
- **13-**Write the descriptions of the tasks you wish to perform on the back of the paper, a single task for each circle. If possible, break the task into segments. For example, if you are working on the circle that represents the trip to the airport, write:
- a.Take car out of the garage – five minutes.
- b. Drive to airport and drop spouse at the terminal – twenty five minutes.
- c. Return home – twenty five minutes.
- d. Return car to the garage – five minutes. Do the same for all the tasks.
- **14-**Put the circles on the physical net and fold it around them. Tie the top with a ribbon, so the papers will not fall out, and suspend it on a hook or a door. It must remain suspended until the tasks are done, or until the seven hours are over.
- **15-**Start with a linear task, which will anchor you. The best one will be the trip to the airport, and for this task no Anunnaki-Ulema powers are used at all. Even though your Conduit is not open, since you have not been trained by a master, it is still there and it can calculate what it needs to do, and how to partially and gradually squeeze the other tasks into the frame of seven hours.
- **16-**When you come back home, you should start the second task, the shopping. While you are shopping, the Conduit will employ a system that will be like two old-fashioned tape recorders working at the same time. One tape recorder is working slowly, about 30 turns per second. The other tape recorder does 1000 turns a second. They do not interfere with each other. While you are shopping, which is represented by the slow tape recorder, the time you are using is slower than the time the Conduit is squeezing in. The Conduit knows how

quickly to "spin" because you have outlined the tasks and the time they take on the circles of paper. This is, therefore, the way the faster tape recorder works.

- **17-**When you come back from your shopping trip, you decide to go to your computer to work on your report. You have to make sure all the physical parts are working properly: The computer is connected to the printer, the paper in the printer is sufficient for printing the entire report, your ink cartridge is fresh, and everything on your desk is in order.
- **18-**Before you start working on your report, unplug the telephone, turn off the TV, make sure nothing is on the stove, and your room's door is locked.
- **19-**Start typing the report.
- **20-**What will happen now will not be entirely clear and understandable to you, because you will be existing, for the duration, on different levels of vibrations.
- **21-**Everything will seem, and actually be, faster than you are accustomed to, including your typing speed.
- **22-**Your body will function normally, but you will not be entirely aware of it, and you will lose your awareness of your physical surrounding.
- **23-**After working for a while, you will feel extremely tired, and without much thinking you will lie down and fall asleep. This is important, because at this time, it is not your normal physical faculties that are in control, but copies of yourself, your doubles, are handling the job. Unless you are a master, it is best to sleep during such occurrences.
- **24-**After a while, and the time for that varies greatly, you will wake up. Naturally, you will return to the computer, feeling again like yourself, and ready to resume your typing.
- **25-**You may be stunned to see that the report of a hundred pages, which you expected to spend hours upon hours preparing, will be neatly stacked by your printer, completely done.
- **26-**When you read it, it will be perfectly clear that it was written by yourself, entirely your work and your style, including your regular mistakes and typos, since the doubles do not edit your work.

- **27**-The only difference is that it was done with supernatural speed.
- **28**-This is a proof positive that you have done the work personally and did not hallucinate these occurrences.

IV. Closing the Energy Center:
You have created a strong field of energy, which now must be closed.
- **1**-Take the net you have suspended, and open it up.
- **2**-Take out the paper circles, and cross out the tasks that have been accomplished.
- **3**-Fold the net and put it in its accustomed place.
- **4**-Throw out the circles.
- **5**-You have closed the energy center, and your tasks are done.

Gomatirach-minzari "Gomu- minzaar": Known as the "Mirror to Alternate Realities."Note: From Ulema Rabbi Mordechai's Kira'at (Reading).
(Excerpts from the books "On the Road to Ultimate Knowledge", and "The Book of Rama-Dosh, both co-authored by Ilil Arbel and M. de Lafayette.) Rabbi Mordechai said: "Building and using the Minzar is risky.
However, if the student reads the instructions carefully and does not deviate from them, it should be a reasonably safe procedure. If you choose to try it, this may be one of the most important lessons you will ever learn, since the benefits, both physical and spiritual, are without equal. Those who are familiar with the concept of the Anunnaki's Miraya would notice a resemblance in the way these tools are used.
However, one should realize that we are not pretending to use the kind of cosmic monitor that is connected, through the Akashic Libraries on Nibiru, to the Akashic Record itself. It is beyond our scope to even conceive how such a tool had ever been created. Nor are we attempting to recreate the kind of Minzar that is used by the Anunnaki-Ulema, who are enlightened beings whose Conduit has been opened.
Most of us possess a Conduit that has not been opened, and the Minzar we recommend is fitted to our level of advancement. Nevertheless, working with the Minzar will open doors that will astound and amaze any student. You will be using the techniques

to create an alternate reality that will allow you to do things you have never imagined are possible. What you are aiming for is a place to which you can retreat at will, a place where you can have many options.

It will be a place of beauty and comfort, and it should allow you opportunities to learn, to create, to invent, to meet compatible people, to connect with animals, to heal, or to simply take a vacation. The place is designed and planned entirely by you, and is brand new. You cannot say "My new alternate reality is exactly like Rome, Italy," because there is a good possibility that the Conduit, confused by this mixed message, will actually take you to Rome, Italy, in our own world.

If this happens, no real harm is done, but no benefit will occur either. You will simply be wandering the streets of another city, not benefiting from the advantages of an alternate reality at all. However, you should certainly take certain elements from places you like, Rome included if that is what you wish, since you are not required to build your new reality in a vacuum.

However, don't limit yourself to one place. You may want to copy a particular art museum from Rome, where you can always indulge in looking at your favorite sculptures and paintings. Then, you might want to add the gorgeous rose garden from the Brooklyn Botanic Gardens in New York City.

A charming old-world train station from somewhere in Eastern Europe might make the place more interesting, with perhaps a touch of the Orient Express, and a sunny Mediterranean beach would not hurt, either. How about a café you liked in Paris, and the cozy little library from your home town, where you used to have so much fun during your childhood and you knew you could find every book that was ever written?

Design the house you would want to live in. It may be an opulent mansion, or on the other hand, some of us would prefer a small, simple, rural-type house with a restful cottage garden. It's all entirely up to you.

Create your new world carefully and don't worry if you change things around as you go along, there is always room for change and development. Did you suddenly remember your trip to China and a wonderful Pagoda you liked? Put it in. Did you enjoy your snorkeling in Australia? Add a barrier reef. One thing should be made entirely clear. Any place you want is allowed, except a place where others are hurt in any way whatsoever, and that includes not only humans, but animals as well. Do not imagine a steak dinner, do not imagine fishing, do not imagine hunting. Don't waste your time imagining the "glories" of wars. Do not imagine a place where you demean your spouse and yourself by having multiple partners.

Do not imagine pornography. Do not imagine a place where you revenge the ills brought on you by people you hate. Your Conduit will not accept any action that can cause pain or even discomfort to any living creatures.

Therefore, if you have any negative intention, you are wasting your time. You can build twenty Minzars, but none of them will take you to such a place. Rather, if you wish to heal from hurts imposed by others, or painful addictions, imagine yourself getting away from all and entering a fresh new world where nothing of this sort exists.

Rest assured that you will never meet anyone who had ever hurt you in your new reality. Do this for a few weeks before you build the Minzar, so the new place is well established in your mind and you can imagine it in seconds.

This is essential because contacting the new reality during the building of the Minzar requires speed, and no one can create a new world for themselves in a few minutes! And most important, don't do it as a chore. This should be a fun, rewarding mental exercise. There is no doubt that you will meet pleasant people in your new reality, but there are those who would also wish to have a guide, or a friend, to introduce them to the new world. This is also possible, and the directions are given below. If this is part of your plan, by all means do the same and imagine the person you wish to contact with.

Don't limit yourself to the kind of person you think you *should* choose. The friend does not have to be a conventional "spiritual guide" which is often described by people who channel entities, such as a Native American guide, an Asian guru, or a guardian angel. The guide can be just about anyone you would like to have as a friend.

I. Prerequisites:
- For seventy-two hours before building the Minzar, and before any subsequent visit to the alternate reality, you must abstain from:
    - Drinking alcohol
    - Using any addictive substance
    - Eating meat.

- - Wearing nail polish
- Do not wear clothes made of polyester.
- Wear white or light colored clothes.
- Imagine only positive conditions (see above for details).

II. Precautions:

- Before starting, remember the full instructions carefully.
- These procedures are for novices, and involve mental transportation only.
- If, however, you become extremely adept, there is a possibility of future physical teleportation. In such event, please exercise some logical restrictions on your activities. For example, people who had heart problems, pregnant women, and those with severe arthritis, asthma, diabetes, should not take the chance of moving physically between realities without consulting first with an Enlightened Master who would advise them on the best way to proceed.
- The Minzar, during building or using, may explode. The explosion is small, and the glass that is used does not shatter or fly around, so you will not be hurt by it. However, if it is built inside your home, or in any confined area, such an occurrence may cause damage to children, pets, furniture, or decorative objects.
- The Minzar must be built in an outdoor location, where the energy that will be released during such an explosion will not cause damage.
- You can build it in your back yard, but if you live in an apartment in the city, you must find an appropriate location where you will be outside, but still have some privacy.
- A woman should not wear loose skirts, flowing dresses, or scarves. For everyone, close-fitting clothes, though not too tight for comfort, are highly recommended.
- Never wear clothes made of polyester.
- Remove any jewelry or metal objects you may be wearing.
- You will be using dry ice. When you handle it, make sure to wear gloves, since direct contact with dry ice will burn your skin.

- You will be using two bowls. Make sure they are not made of metal.
- When you cut the dry ice, be sure to place it in the dry bowl. Never mix dry ice and water, this can cause serious injury.

III. Equipment and Supplies:
The supplies required to build the Minzar are readily available. You will need:

- Laminated glass, two feet by two feet, with rounded, smooth edges. Laminated glass is made of two layers of glass, and it does not shatter into sharp-edged slivers when it breaks. It is the safest glass you can use. Have the store cut it for you to the right dimensions.
- A few pieces of charcoal
- A role of aluminum foil
- A very small quantity of dry ice. You will only need a small cube, approximately the size of a dice.
- Two very thin pieces of wire, each three feet in length
- Two iron nails
- A Magnet
- Two plastic or glass bowls that would contain sixteen ounces of liquid each. Never use metal.
- Lumber, enough to build a two feet by two feet base, two inches height
- Wood glue
- Adhesive spray
- Fabric glue
- Small finishing nails
- A small hammer
- Water.
- A sheet of white linen, large enough to create four panels that you will use to surround yourself as you work with the Minzar. This sheet should be made of flame-retardant fabric, or if you cannot find such a sheet, spray your linen with flame-retardant spray.
- Four Pieces of cardboard, six feet by two feet.

IV. Building the Minzar:
- Magnetize the iron nails by placing them next to the magnet for a few hours.
- Build a wooden base. It should be a simple box, two feet by two feet, and two inches tall. Use the wood glue and the finishing nails to make it steady.
- Fold each piece of cardboard vertically, ending with a small pyramid measuring three feet by two feet. Make all four can stand up steadily.
- With the fabric glue, attach four panels from the white linen sheet to the cardboard pyramids.
- Rub the coal on one side of the glass, until it covers the surface with a thin black film. Use the adhesive glue spray to stabilize the film. Allow to dry thoroughly.
- From the aluminum foil, cut seven ribbons. Each should be a little less than one inch in width. Six of the ribbons should be exactly two feet long, and the seventh should be two inches longer.
- Take four ribbons, not including the longer one, and glue them to the coal covered side of the glass. They should be placed with equal distance between them and from the edges, creating five equal sized spaces where the coal dust will be visible.
- Take the remaining three ribbons. They should be glued on top of the four ribbons, but in ninety degrees to them, creating a grid. The longer ribbon should be glued in the middle of the box, with an inch extending on each side. The others should be glued with equal distance between the middle ribbon and the edges, creating four spaces. The grid will thus be made of a square spaces between the ribbons.
- Use the extra ribbon that is extending from both sides to attach the wires. Each wire will be extending vertically from the box.
- Place the glass on the wooden base, coal and ribbon side down, and clean side up. Make sure the glass and the base are squared and the edges are perfectly aligned.
- To each wire, attach one of the magnetized nails you have prepared in advance.

- Arrange the panels around the box. There should be one on three sides, and the fourth one will be placed behind you.
- Pour the water into one of the bowls, and place one of the nails into it. The wire that is attached to this nail must be fully stretched.
- Cut the dry ice, wearing gloves, into a dice-sized cube. Place it in the dry, empty bowl. Remember never to mix dry ice and water! That wire should be closer to the glass than the one that is touching the water, so bend it slightly.
- The dry ice will produce some smoke. That is normal, it is an effect that is often used for theatrical production, and it will not hurt you.
- Sit in front of the glass box, put the fourth panel behind you, and close your eyes.

V. Contacting the Alternate Realities:
- Close your eyes and visualize a green, virgin land, a place no one has ever seen before.
- Imagine, dream, and think about the land you have been visualizing for the past few weeks. You are bringing the things you love and want most, the good things that you wish to see in your life, to the green land. You are creating a new earth, the way you want it.
- There are people in the new place.
- You must build places for them, streets, houses, a wonderful city or countryside, exactly the way you want it. Working as fast as you can, and with your eyes still closed, in a few minutes you will sense smoke coming from your left side. It will not rise high, but remain rather low, and it will creep close to the glass. Realize that even though your eyes are closed, you will actually see the smoke.
- When you are sure you are seeing the smoke, open your eyes.
- Put both your hands on the glass, with your fingers spread out.
- Concentrate your gaze on the spaces between the fingers. Bring to mind all the beautiful things you imagined in

the new land, and place them in the spaces between the fingers.
- Start alternating your concentration between the tips of your fingers and the spaces between the fingers. Continue for about five minutes.
- You will notice that the tips of your fingers will produce light, in the form of sparks. There will be no physical sensation caused by these sparks.
- Slide your hands closer to your body until they are about an inch or two from your body.
- Put your hands on the edges of the glass, each on one side.
- Look down into the bottom of the Minzar. You will notice that the color of the aluminum ribbons has changed, and that the charcoal film looks as smooth as a marble. The glass has turned into a black mirror, and a line of light will vibrate on the black surface.
- You will begin to see the things you have imagined as miniatures in the black mirror.
- Some will look proportional and organized. Others will be out of proportion. They will be moving and shifting.
- You may have created a person to function as a friend and a guide. If you did so, look for that person in the Minzar.
- You will soon find him or her, so try to increase the size of the person. In a few seconds, the person will acquire dimension, proportion, and personality, and will appear as real, in or out of the Minzar.
- You will establish a true rapport with him or her, though you may not quite understand the nature of the rapport.
- If what you imagined is a country, or a place, or a house rather than a person, you will develop the connection to it so that you will be able to escape to this place at will. Many students prefer creating such a place, since, as it will most likely to have people in it, will combine the advantages of both.
- In the future, you will not need to build a second Minzar, or even use the many steps of preparations to envision the person or the place you have created.
- They will be stored in your brain. The act of building the Minzar was meant to trigger one of the Conduit faculties

in the brain. A rudimentary one by comparison to what the Anunnaki-Ulema can do, but of great benefit none the less. You could not, for example, simply buy a ready-made black mirror, and work with it. You must follow the step-by-step the creation of the Minzar to achieve the effect.

- It will be a good idea to throw out the unnecessary equipment, such as the nails, the bowls, etc., but keep the Minzar, which has turned into a beautiful black mirror, as a stimulus for the activity.
- You can go into the new country anytime you wish. It is a physical place, located in a different dimension, but just as real as this one.
- When you go there, you can spend months in that time frame, while here on earth only a few minutes will pass. That is because the Conduit allows you to duplicate yourself, to create a double, and time is different in other dimensions.
- What you can do there is limitless. You can simply rest and enjoy a place that will never hurt you, a vacation from the trials and tribulations in the here and now. Or, perhaps, you wish to create something.
- Let's say you want to write a screenplay, and can never find the time or the leisure to do it here. Well, you can go to your special place for the duration of the time you need for writing this screenplay, and come back to your present existence after a few seconds of leaving it.
- The advantage will be that you have written the play and it is all there in your memory, one hundred percent of it. All you will need is the short time needed to type it.
- Or perhaps you are not well, and you would like to see the doctors and the hospitals you have created at this new environment.
- It is quite likely that they may have a cure to at least some ailments – it won't hurt to try. Possibly you wanted to build a magnificent library, containing an enormous number of books. By all means, this is a wonderful experiment, with one added bonus.
- When you are at this library, make a note of certain titles/authors which you have never heard of before.

Then, when you are back home, ask a librarian, or check the Internet, to see if such titles/authors exist.
- If they do, it would be a proof that you have not been hallucinating! Or perhaps you would like to try a new career, see how it feels to become a teacher, or a singer, or a trapeze artist. Why not try it? You are the best judge on what you wish to accomplish!

VI. Subsequent Visits to the Alternate Realities:
After the initial visit to the alternate reality, you will no longer need to use the Minzar. As mentioned before, some students find it easier to look at the Minzar for a while before attempting the visit, but it is not entirely necessary.
- The best time to visit is your usual bed time. Before you go to sleep, just lie down on your bed. Generally, it is best to lie on your right side, to avoid pressure on the heart.
- Close your eyes. Think about the place you want to visit. Draw as clear a picture of it in your mind as you can. At this point, remember the way your hands were placed on the Minzar, and imagine yourself behind your fingers.
- Tell yourself the first activity you wish to perform during your visit.
- For seven to ten seconds, do not think at all. Make your mind completely blank.
- Do not be startled – amazing things will begin to happen now. Images will float before your eyes, you will hear sounds, or noises. This is called "The buzzing of the mind."
- At this moment, the preliminary rapport is established between the necessary cell in your Conduit and your double in the alternate reality. The cell will zoom you there and your double will be your guide. In other words, the cell acts as your vessel, and the double as the pilot.
- As soon as you arrive, the double will stop all activities and instantly merge with you. Your visit has begun.

VII. Benefits and Advantages:
Beside the pleasure and learning experiences that you gain through your trips to the alternate reality, there are several

concrete advantages that will manifest themselves very soon in your normal reality.
- You will be less tense or nervous.
- You will gradually lose any phobia that might have tormented you for many years, perhaps all your life.
- Your physical health will improve.
- You will be able to work efficiently, since you will bring with you some very important creations, plans, or thoughts from your alternate reality. Such products or services will be performed in much greater speed since they have been "rehearsed" in the alternate reality.
- You can learn languages with surprising speed since you can actually learn them first in the alternate reality, and the memory is retained. That applies to other skills, such as computer skills, art, music, and many others.
- You will put every moment to good advantage. If you hate waiting in line, or sitting in the doctor's office, or listening to your boss droning on and on while of course you cannot put a stop to the conversation, just hop to the new reality for a few minutes, and do something fun or creative there. Of course, for these few moments you will be out of touch with your earth body, but you will be recalled back quickly as soon as needed. Obviously, using this quick "hop" you will never be bored again, ever. To complement this activity, it is advisable to always carry a notepad and a pen in case you wish to quickly record an experience.

VIII. Returning to Your Regular Reality on Earth:
We must note that there is never any need for fear. Some people are concerned that the body that they have left on earth when visiting their alternate reality might be exposed to harm, perhaps even attacked.
There is no reason for such fear. First of all, with the exception of the first time, when you originally build the Minzar, you will usually do it in the privacy of your own bedroom, and alone. Second, no matter how long you will spend in your alternate reality, you will return to your body seconds after you left it in our reality here, since time flows very differently in the alternate reality, and the Conduit knows how to handle it. The only thing you should be concerned about is not to come back into the body

too quickly. If you panic suddenly and zoom into your body, you may harm it by this speed. You are perfectly safe, so come back easily and slowly.

The best procedure for a beginner is to spend the time and enjoy the stay in the alternate reality without worrying about coming back. The first few times would not take long, since you are so new at it, anyway.

After a while, your stays will be extended. In both cases, after what seems to be minutes, hours, days, or months, since it really does not matter how long you are there, suddenly you will remember that you left your body behind. For a few seconds, you are not sure which part of you is real, and it may create the sense of fear discussed above.

Remember there is nothing to fear, your Conduit is in control, and it knows what it is doing. So when this moment arrives, allow yourself to relax, and in seconds you will be aware that you are back in the presence of your normal earth body.

Do not rush, and do not bunch yourself quickly into the body from either side.

Instead, help your Conduit by hovering horizontally right above your body, and then settling peacefully into it. Most likely that will be followed by a few minutes sleep, after which you will wake up refreshed and in complete memory of your activities in the alternate reality."

Gubada-Ari: Term referring to the Anunnaki-Ulema "Triangle of Life", and how to apply the value of the "Triangle" shape to health, success, and peace of mind. Note: From Ulema Rabbi Mordechai's Kira'at (Reading). (Excerpts from the books "On the Road to Ultimate Knowledge", and "The Book of Rama-Dosh, both co-authored by Ilil Arbel and M. de Lafayette.)

Ulema Rabbi Mordechai explained the concept, importance, and practical use. He said (Verbatim):

How this technique will enhance your life:

With the help of the triangle, you will able to find the perfect areas on earth where your health, success, and peace of mind will be at their optimum. You can work it on a large scale and find out the best countries to live in, or on a small scale, which would give you the best neighbourhoods in your own city or county.

I. Synopsis of the Theory:
- There are lines of energy spinning around the world. In this exercise, we will concentrate on the lines that are revealed by the use of the triangle.
- The energy flows in currents, both negative and positive, mostly underground, traversing the globe.
- Those who live above the positive lines, will have good health, success, and peace of mind. Those who live above the negative lines, will have bad health, lack of success, and will experience mind turmoil.
- The meaning of life is based on the fact that life is, in itself, a triangle.
- One corner of the triangle represents health.
- The second represents success.
- The third represents peace of mind.
- You find meaning by placing the triangle you are about to draw on the world.
- The student might ask, where do I put the triangle? How do I choose the original location? The answer is, you put the triangle wherever you are.
- The student might ask, what if I change locations? The answer is, this technique is working within the dictates of the moment. Wherever you are, the triangle follows. Change it as many times in life as you need. It always works.

II. Materials:
- This lesson can be accomplished with two different props.
- You can use a globe, or a flat map of the world.
- A globe gives a more precise directions, but it is expensive and sometimes hard to get.
- The student may instead use a flat map of the world.
- It is not as precise, but the distortion is so slight that it does not signify, and it is cheaper and readily available.
- If you are using a map, you will need lightweight paper which is somewhat transparent, a pen, a ruler, and a pair of scissors.

- If you are using a globe, you will need plastic wrap, the kind that is used to wrap sandwiches or leftovers in the kitchen, since it will adhere easily to a globe.
- You will also need a magic marker that can write on this material, a ruler, and a pair of scissors.

III. The Technique:
- The drawings below show how the double triangle, or the six-pointed star, was created.
- To be most effective, an individual exercise should be used separately for Health, Success, and Peace of Mind.
- As you copy the template below, simply change the word on top for each exercise.

"Triangle A" was drawn as an equilateral triangle.

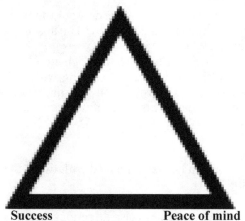

Success   **Peace of mind**
Figure 1: Triangle A

*** *** ***

"Triangle B" was drawn by extending the lines on top of triangle A, and then closing these lines and thus creating a second triangle of the same size exactly.

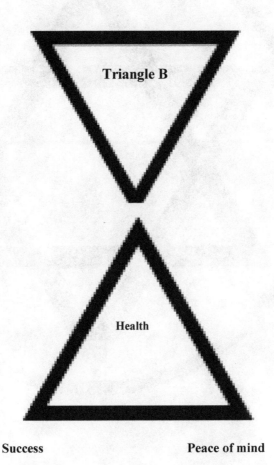

Figure 2: Triangle on the top is triangle B.
"Triangle A" was moved up and centered exactly on Triangle B. By doing this, we have created a six pointed star.
We have numbered the four small triangles created on the sides of the star, as 1, 2, 3, and 4.

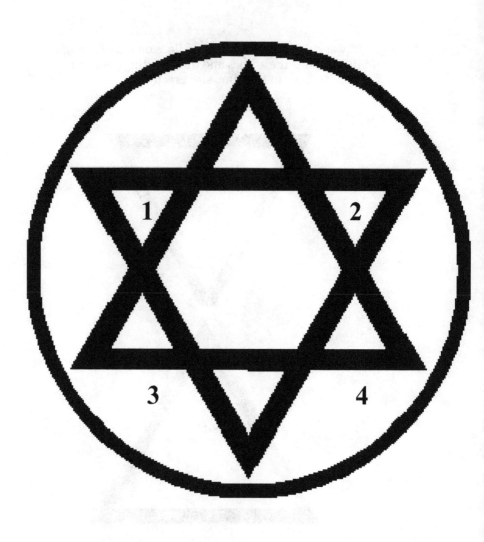

Figure 3: The six pointed star

- Copy the template of the star on transparent paper if you are using a map, or on the plastic wrap if you are using a globe.
- Place the center of the star precisely on the location of the place you are living in now, at this very moment.
- All countries located inside these four small triangles are good for your health. Should you have a health issue, or a desire to live in the more healthy places, these are your choices

*** *** ***

The Anunnaki-Ulema Triangle is a complex concept. Perhaps, to better understand its essence, revisiting Ulema Mordechai, and listening to him explaining to his student Germain Lumiere is a must. Excerpt from his dialogue with Lumiere, reproduced from the books "On the Road to Ultimate Knowledge", and "The Book of Rama-Dosh, both co-authored by Ilil Arbel and M. de Lafayette.) "We are going to apply the value of the triangle shape to real life and to the organization we call the Pères du Triangle. I am not sure if you are aware of it, but there are six Triangles on earth. Actually, they rule the earth." Lumiere: "Are they political? Secret? Are they part of governments' agencies?" I asked.
"They are more important, far more so, than mere governments. Can you define for me what are the most important things in life?"
"Life itself?" I said.
"Yes, this is right within itself, but it does not answer the question." I was annoyed. Here we go again, I said to myself. I am arguing with an old Jewish Rabbi. They always go round and round, using semantics that get you nowhere. "How can I be right and wrong at the same time?" I asked.
"Well, we will go about it in a different way," said Rabbi Mordechai. "What is the meaning of life on earth?"
"Family? Friends?" I said, knowing full well that he will argue again, and I was right.
"Family and friends make our life meaningful, of course," said Rabbi Mordechai, but they are not the meaning of life. The meaning of life is based on the fact that life is, in itself, a triangle. One corner of the triangle represents health.
The second represents success.

The third represents peace of mind. Visualize it like that." And he demonstrated by joining his two thumbs and his two forefingers, creating a triangle. "You find meaning by placing this triangle on the world." He leaned his hands on the large globe. "But the all important thing is to find the right spot to put the triangle on."
"I am not sure I follow," I said, dubiously.
"So let's demonstrate it with some props," said Rabbi Mordechai. He gave me paper, pencil, a ruler, and a pair of scissors. "Now," he said, "draw and cut a more or less equilateral triangle from this peace of paper." I did, trying my best to make an exact drawing, and cut it carefully.
"Now," he said, "put it anywhere on the globe."
I took the paper, and feeling like a fool tried to place the paper on the globe, knowing that it will fall off since I used no glue. Of course it fell, several times, until Rabbi Mordechai smiled rather cynically, an expression I have never seen on his face before. "Put it on again," he said, giving the globe a piercing look. I did, and the paper stuck to the globe. Another trick, I thought. I was tired of tricks.
"Spin the globe," he said. The triangle stuck and the globe was spinning.
"As this is happening," said Rabbi Mordechai, "realize that if the lines of the triangle were somehow continued, they would represent lines of energy around the world. Let's concentrate on the lines that occur when you extend the Health corner at the top of the triangle.
This energy flows in currents, both negative and positive, mostly underground, traversing the globe." This was beginning to make sense to me. I span the globe again, the paper stuck, and I tried to imagine the continued lines that would follow the entire world. I was beginning to see the pattern.
"Those who live above the positive lines, will have good health. Those who live above the negative lines, will have bad health. But let's elaborate a little. Look at the drawing I am about to make."
He drew a triangle, wrote Health on the top of it, and said, "This is Triangle A."
Then, he extended the lines. "Close these lines and thus create a second triangle of the same size exactly, which we call Triangle B. Everything inside Triangle B will have good health. Now, make a copy, of an exact size, of Triangle A. Move it up and center it exactly on Triangle B. By doing this, you have created a six pointed Star of David."

By now I realized we were not doing any tricks, but studying a most fascinating and helpful technique. "How do we proceed?" I asked, poring over the drawing.
"We will number the four small triangles created on the sides of the Star of David
1, 2, 3, and 4. All countries located inside these four small triangles are good for health. Should you have a health issue, or a desire to live in the more healthy places, these are your choices."
"So I imagine that you can do the same for Success and Peace of Mind, to find the best of each quality?"
"Correct," said Rabbi Mordechai.
"Ah, but there still one problem here. Where do I put the triangle? How do I choose the original location? " I asked.
Rabbi Mordechai laughed. "For once, son, I encourage you to consider yourself the center of the world. You put the triangle wherever you are."
"However, Rabbi Mordechai, another question remains. At this moment I am in Budapest. I put the triangle on the map of Hungary and learn of my best locations. But next week, or next month, I am going back to France. Then, should I put it on the map of France?"
"Yes, of course," said Rabbi Mordechai.
"This technique is working within the dictates of the moment. Wherever you are, the triangle follows. And it always works."
"I am a little surprised to see the Star of David involved in Ulema teachings," I said.
"Not at all," said Rabbi Mordechai. "You must realize that the Kabbalists share many of the Ulema techniques. There is much more to it, as this is only one of the seven great secrets of the Star of David," said Rabbi Mordechai. "The Kabbalists have been using it to great advantage for centuries."
"But the Triangle is used by the Pères du Triangle, so it is a universal symbol," I said.
"Good point. As you can imagine, the presence of the Star of David caused the usual Anti Semitic comments that the Jews are ruling the financial world. But this is sheer nonsense. The Pères du Triangle include people from all religions and nations, and they have very little affiliation to either.
The Star of David, even though it signifies in Judaism and is placed on the flag of the state of Israel, is entirely universal and many scholars claim its origin is Anunnaki." Indeed, so much of

the Ulema knowledge comes from the Anunnaki, that it did not surprise me."

Gudinh: To move around; the act or the attempt of bringing things (even thoughts) together and creating a virtual three dimensional reality from assembling collected thoughts, desires and ideas, and projecting them over a mirror serving as a catalyst. This mirror is called Miraya in Ana'kh language.

Hag-Addar: The right to enter a palace; a metaphoric expression for an Ulemite spiritual initiation, sometimes referred to as the 18th degree ritual ceremony. According to Ulema Govinda, during this ceremony, the adept is taught the secrets of the origin of the creation, and the invisible dimensions that co-exist with the physical world. In addition, a multitude of techniques are revealed to the initiated, that allow him/her to acquire extraordinary powers, such as teleportation, Tay Al Ard, Firasa, dematerialization, and psychotelemetry.

Hamnika-mekhakeh: Kha in mekhakeh is pronounced as "Jo" in Jose in Spanish. Grids used by Anunnaki-Ulema as calendar to find the lucky days and the lucky hours in a person's life.

Hamnika-mekhakeh-ilmu: The technique of using the Hamnika-mekhakeh. (Source: The book "On the Road to Ultimate Knowledge", co-authored by Ilil Arbel and Maximillien de Lafayette.)

I. Synopsis of the concept:
Humans follow certain calendars. The most common one is the Gregorian Calendar, which is a reflection of the Christian faith. It is younger than the Muslim calendar, which in turn, is younger than the Jewish calendar. All of these are considerably younger than the Anunnaki calendar, which is the only one used by the Anunnaki-Ulema. The Anunnaki-Ulema reject the idea that the week consists of seven days. Their week consists of four days, corresponding to certain days of our week. These are the only days to use in this technique, and the other three days in our week should not be calculated upon.

II. The Ulema-Anunnaki days are:
- Day 1: Thilta (Tuesday)
- Day 2: Araba (Wednesday)

- Day 3: Jema (Friday)
- Day 4: Saba (Saturday).

The importance of these days is the relationship between the person and the hours in each day. Using the calendar of the Anunnaki-Ulema, each person can find the luckiest hour of his or her week, according to the Book of Ramadosh (Rama-Dosh). Ulema Rabbi Mordachai said: "You might feel that one hour a week is not sufficient for anyone's needs. It might also not improve your luck at work if it occurs, say, at two o'clock in the morning each Saturday.

This predicament can be easily resolved by performing another technique, Time Manipulation, on that exact hour. The time that will be added to your life under such circumstances will be as lucky as the original hour, and your chances of success will be vastly improved."

The Anunnaki-Ulema teachers highly recommend performing a combination of techniques, since each enhances the other considerably.

IIL The calendars' grids:
A couple of questions might arise as you work with this technique. First, are all people with the same number of letters in their name share a lucky hour? Yes, indeed they would. There are only sixteen grid lines to represent millions of people each. And this leads to an interesting discovery. The numbers of letters in people's names represent a certain harmony that exisst between them. For example, if you wish to approach someone in high places for a favor, finding that he or she shares the number of letters and the lucky hour will enhance your chances. Always send your request to him or her during the lucky hour, calling on the phone, using your e-mail, or placing a written letter in the mailbox.

IV. The use of a language:
Another question is the issue of languages. What if your name is written with four letters in America, where you live, but with five letters in your native language? The answer is simple. Always use your native language, the language that you were first aware of your name in, in your grid. It will be much more accurate and certainly more powerful. An important fact to add is that this technique is simple, but it can be enhanced in many ways by

subtle variations. Adding those variations extends the knowledge of how time and space is related to luck and success, and how to fine tune the process. But even in this straightforward version, the technique is incredibly powerful, so much so that it may change your life completely, always for the better.

> **Tip:** If any added numbers are higher than one digit, always add the numbers and use the result. For example, if instead of 3+1+1+1= 6 you will find yourself with, say, 4+7+7+7=25, add 2+5 and use the result, namely 7. If you have 40+41+42+43=126, add 1+2+6=9.

V. The preparation and use of the grids:
The first step is to prepare a grid of sixteen squares, like the one below.

In the next step, you will establish the calendar of the week, by writing them in this specific order.

Grid 1: Calendar of the Week

Day 1	Day 2	Day 3	Day 4
Day 2	Day 3	Day 4	Day 1
Day 3	Day 4	Day 1	Day 2
Day 4	Day 1	Day 2	Day 3

- **1**-In the next step, you will establish the calendar of your name. Let's say your name is Suzan.
- **2**-You will write your name in the squares, but you must write from right to left, the way they did in many ancient languages, including Ana'kh.
- **3**-Then, you follow, still from right to left, with the number of the days, 1, 2, 3, 4.

Grid 2: Calendar of Your Name

A	Z	U	A
3	2	1	N
Z	U	S	4
2	1	N	A

*** *** ***

Calendar of your Lucky Hour

- **1**-In the next step, you will establish the calendar of your lucky hour.
- **2**-Look at the two squares above.
- **3**-Try to find the one square that has the same number in both drawings.
- **4**-When you compare each square, you will see that the second square in the last raw has the #1 in it.
- **5**-Fill in the number of the days in the first row, the way it appeared in the first grid.

Therefore, Suzan's lucky hour will occur during the second day. (If more than one square presents the same number, add the numbers.)

*** *** ***

Grid 3: Calendar of your lucky hour

Day 1	Day 2	Day 3	Day 4
			1

In the next step, we shall start our calculations.
- **1**-Keep the first row as is.
- **2**-fill the rest of the grid with the number 1.
- **3**-In each column, you will now subtract the three #1 from the day in the first row. 1-1-1-1= -2; 2-1-1-1= -1; 3-1-1-1= 0; 4-1-1-1 = 1

Grid 4

Day 1	Day 2	Day 3	Day 4
1	1	1	1
1	1	1	1
1	1	1	1
-2	-1	0	1

- **4**-We will now add the number we have calculated. (-2) + (-1) + 0 + 1 = (-2)
- **5**-We continue our calculations by using the number we have achieved, -2, as a filler in the grid below, in three rows under the basic days row on top.
- **6**-Then, we will calculate the values of the columns the way we have done in the previous grid.

Grid 5

Day 1	Day 2	Day 3	Day 4
-2	-2	-2	-2
-2	-2	-2	-2
-2	-2	-2	-2
-5	-4	-3	-2

- **7**-We will add these numbers: (-5) + (-4) + (-3) + (-2) = -14
- **8**-We will combine the individual numbers comprising the number fourteen by adding them: 1+ 4 = 5
- **9**-We will add these two numbers. (-14) + 5 = -9

*** *** ***

In the next step:
- **1**-Return to the first grid, displaying the calendar of the week.
- **2**-Starting on the second row, count the squares, going from right to left, nine times.
- **3**-You will reach Day 3.
- **4**-This establishes that your lucky hour will occur on Friday, the third day of the Anunnaki week.
- **5**-To establish the hour, go back to Grid 4, and look at the row that expresses Day 3.
- **6**-Add the numbers: 3 + 1 + 1 + 1 = 6
- **7**-Calculate: (-9) - (+6) = -3
- **8**-To establish the hour within the 24 hours in each day subtract, 24 – 3 = 21.

21 is 9 P.M.
Therefore, Suzan's luckiest hour of the week occurs at nine o'clock in the evening of each Friday. (Source: The book "On the Road to Ultimate Knowledge", co-authored by Ilil Arbel and Maximillien de Lafayette.)

Hamsha-uduri "Dudurisar": Term for the act or ability to rethink and examine past events in your life, change them, and in doing so, you create for yourself a new life and new opportunities. To a certain degree, and in a sense, it is like revisiting your past, and changing unpleasant events, decisions, choices, and related matters that put you where you are today.

I. The Concept:
The honorable Ulema spoke:

- 1-You are not totally satisfied or happy in your current situation, in what you are doing, and where you are today. You wish you had a different life, a better job, more opportunities, less troubles and worries.
But there is nothing you can do about it, because the past is the past, and nobody can change the past. It is done.
- 2-True, it is very true, you cannot change your past, because it was Maktoub (Written) in the pages of your fate and destiny book, the day you were born.
- 3-Besides, changing or altering the past creates global chaos, confusion, and dysfunctional order among people, communities and nations.
- 4-But you don't have to change your whole past (All events) to become happier, more successful, and put an end to your misery and tough times.
- 5-All what you need to do is to change the part or segments (Days, months, years, places, decisions) that have created hardship, misfortunes and mishaps in your current life; and this is permissible, possible, ethical and healthy.
- 6-What part of your life you wish to change or alter, how to do it, and the reason(s) for doing it are the paramount questions you have to ask yourself very honestly.
- 7-Once you have decided to alter a portion or a segment of your past, and replace this specific segment with new wishes and decisions, you cannot change, modify or alter the changes you have made.
- 8-You have to live with it, and you will not have another chance.
- 9-Something else you should think about (and you should never forget it): Your family, your children, your

current obligations, your commitment to others, and all the promises you made to people. Because, once you have changed a major part of your life (Past or present), all these things that you are dealing with right now will change too, and you might not like it, or perhaps you might? Everything depends on your intentions, needs and desires.

- **10**-Avoid at all costs and at all time selfishness, greed, escaping from responsibilities, and vicious intentions. No, you can't go back in time and kill all those people who did you wrong, or hurt the woman who turned you down.
- **11**-You go back in time/space for one reason: Changing unpleasant events and decisions that unfairly have caused you grief, failures, pain, and unhappiness.
- **12**-So the keyword here is "Unfairly". But how would you know what is fair and what is unfair, as far as your past, your job, your luck, your fortune, your bank account, you relationships, promotion in your career or profession, health conditions, peace of mind, and success are concerned?
- **13**-You will, once you have crossed the frontiers of the present-past sphere.
- **11**-You will not be able to lie to yourself. Once you are behind that thin curtain, you become a different person; un-materialistic, honest, sincere, simple, and wise.
- **14**-The world you will enter is as real as the one you live in right now.
- **15**-However, it is much much bigger, prettier, serene, meaningful, and confusing at the same time, because it has different purposes, unlimited borders and levels, it is ageless, multidimensional, and contains places, dates, times, people, unfamiliar creatures, humans and non humans (Women, men, children, even animals and strange non-physical creatures) you have not met yet in your life, because they either belong to the future, and/or to different, multiple, vibrational, parallel worlds.

II. The Technique: It works like this:

Note: This might frustrate you, because it requires a great deal of patience. Without patience, a strong will, and perseverance, you will not achieve a thing. This technique works.
It is up to you to make it work for you.

- **1**-First of all, you have to understand that you cannot revisit and change your past at will, or as you please. There are rules you must follow. And these rules are closely connected to time, the place where you live, your intentions, and the specific events you wish to change.
- **2**-For instance, you cannot return in time/space to take revenge, to kill a person you dislike, or prevent that person from evolving or competing with you.
- **3**-There are limits to what you can do. You cannot return in non-linear time/space and create wars or bring destruction to people and communities. Changing a part or a segment from your past is limited to your sphere.
- **4**-Sphere means your very personal life, your personal actions, your personal events; things that are closely and exclusively related to you.
- **5**-You have no powers over others.
- **6**-Also you have to understand that you cannot return to the past any day you want. There is a calendar or more precisely a time table you should fit it. This means, there are days and hours that are open to you; these days and these hours allow you to zoom back into your past.
- **7**-Consequently, you must find out what are these days and these hours. You have to consider yourself as if you were an airplane.
- **8**-An airplane cannot take off or land without a flight schedule, and an authorization. Consequently, you have to schedule your trip to the past according to a trip schedule.
- **9**-The trip schedule is decided upon by many factors. The two most important factors are:
- a-The activation of your conduit;
- b-The most suitable hour for your trip; this happens when your Double, or at least your astral copy is in a perfect synchronization with the Ba'ab's opening. Possibly, you are getting confused with all these words and conditions. I will try to simplify the matter for you.

- **10**-To find out what is the most suitable hour to schedule your trip to the past is to do this: All the letters of the alphabet have a numerical value in the Ulemite literature, and were explained in the Book of Rama-Dosh). These numerical values are as follows:
- A numerical value: 1.
- B numerical value: 3.
- C numerical value: 5.
- D numerical value: 7.
- E numerical value: 11.
- F numerical value: 15.
- G numerical value: 16.
- H numerical value: 21.
- I numerical value: 32.
- J numerical value: 39.
- K numerical value: 41.
- L numerical value: 42.
- M numerical value: 44.
- N numerical value: 49.
- O numerical value: 56.
- P numerical value: 58.
- Q numerical value: 62.
- R numerical value: 75.
- S numerical value: 81.
- T numerical value: 83.
- U numerical value: 89.
- V numerical value: 95.
- W numerical value: 98.
- X numerical value: 102.
- Y numerical value: 111.
- Z numerical value: 126.
- **11**-Let's assume your name is Kamil. Now, let's find out the numerical value of Kamil.
- **K=41**. Add 4 + 1=5. The number 5 is the numerical value of K.
- **A=1**. A is a single digit. The number 1 is the numerical value of A.
- **M=44**. Add 4 + 4=8. The number 8 is the numerical value of M.

- **I=32**. Add 3 + 2=5. The number 5 is the numerical value of I.
- **L=42**. Add 4 + 2=6. The number 6 is the numerical value of L.
- **12**-Now, let's find out the complete numerical value of the whole name of Kamil. Add the numerical values of all the letters of the word Kamil.
- **13**-We have here: 5+1+8+5+6=27.
- **14**-The numerical value of Kamil is 27. Now add 2+7=9. The number 9 represents the 9th hour on Kamil's trip schedule. Always use night hours. This means that you can't zoom into your past at a different hour. Just tell yourself you got to be at the gate at 9 o'clock, exactly as you do when you take a regular flight at any airport. You have to be on time.
- **14**-Now, we have to find out the day of the trip (Zooming into the past.) Only on this day you will be able to do so.
- **15**-Take the "Triangle" you have used in practicing the "Gubada-Ari" technique.
- **16**-Place the Triangle on the globe. If you have done this previously and found out your best spots on the globe, then, take the number corresponding to that particular spot. Remember there are 4 small triangles, and each triangle has a number on it (From 1 to 4 on the Six Pointed Star.)
- **17**-Let's assume that the corresponding small triangle's number is 3. What does 3 represent? Or any of the 4 numbers of the four small triangles in any case?
- **18**-On Earth, the week consists or 7 days. In the Anunnaki-Ulema Falak, there are no weeks, no days and no time, because time is equated differently beyond the third dimension. Instead, the non-terrestrial beings have four spatial sequences, each representing four different spheres beyond the fourth dimension. Don't worry about these spheres. For now learn this: The number 3 you found on the globe means the third day of the Anunnaki-Ulema calendar.
- **19**-The Anunnaki-Ulema calendar consists of the following: 4 days (So to speak). And the four days are (In a figurative speech) 1-Tuesday; 2-Wednesday; 3-Friday; 4-Saturday. Forget all the other days.

- **20**-This means that your zooming day is Friday. And the departure is at 9 (See bullet#14).
- **21**-Now, we know the day and the hour. What's next?
- **22**-Now, we have to find the gate (Ba'ab).
- **23**-The gate is usually very close to you. This gate is not the "Spatial Ba'ab", but your mental launching pad.
- **24**-The most practical launching pad is any place where nobody can interfere with this technique.
- **25**-Use your private room, and make sure that nobody is around.
- **26**-Place the Triangle on your desk, or on any solid and stable surface.
- **27**-Sit in a comfortable chair behind your desk, close enough to place both hands parallel to the triangle.
- **28**-Take a deep and long breath, approximately four or five seconds. Repeat this three times.
- **29**-In your mind, project two lines exiting from the top corner of the triangle. In other words, let the lines of the both sides (left and right) of the triangle continue outside the triangle leaving from the top corner.
- **30**-In your mind, create a new triangle starting from the top of the triangle you already have on the desk, and just formed by the extension of the two lines you have mentally created.
- **31**-Now you are looking at this new triangle sitting on the top of the other triangle. We are going to call it "Mira".
- **32**-At each corner of the Mira, you are going now to put something.
- **33**-For example, at the base, which is upside down (left side), choose the year you want to visit.
- **34**-At the base, which is upside down (right side), pick up the event you want to revisit. By event, I mean something you have done, such as a decision, a job you had, a responsibility you have assumed, a trip you made, an encounter you had. In other words, the thing you want to change or alter from your past.
- **35**-At the lower corner of Mira which is joining your triangle on the desk, put the "new thing" or "new wish" you want to use to change or alter something you have

done. Remember: You have already chosen what you want to change; it is already placed at the right side (upside down) of the triangle (See bullet#34).
- **36**-Now, everything is in place. You have on the left side, the year you want to visit; on the right side, you have the event or the thing you want to change, and on the reversed corner of the triangle you have the new thing you want to bring to your past so it could change or alter the things you want to remove from your past.
- **37**-Your final mental projection is this: Draw in the triangle a circle, and make it spin or rotate in your mind.
- **38**-Let the circle spin as fast as possible.
- **39**-Keep telling your mind to order the circle to spin. And concentrate on the rotation of the circle as strong as possible. Stay in this mode as long as it takes until you see the circle increasing in size.
- **40**-When you begin to feel that the circle is getting big enough to contain all the corners of the triangle, at this very moment, and in your mind, push out of the circle all the contents of the right side corner, and immediately refill it with your wishes.
- **41**-Bring the refill to the left side corner.
- **42**-Anchor it in the chosen year.
- **43**-In your mind, close the right side corner.
- **44**-In your mind, close the bottom corner of the triangle.
- **45**-In your mind, stay with the year you are visiting now.
- **46**-You are now seeing lots of things; scenes from your past, people you knew, it depends where you are, in other words, your past as it happened at that place and during that year has been reconstructed. Not hundred per cent physically, but holographically, and in this holographic projection, the essence, fabric and DNA of everything that happened are brought back to life in a different dimension, as real as the one you came from.
- **47**-Remember, nothing is completely destroyed in the world. Nothing comes to a total and permanent end. Everything is transformed, retransformed, and quite often recycled. You can recreate the original fabric of everything that happened in your past by recreating the DNA sequence of past events, in revisiting the parallel dimension that has never lost the print and copy of

original events. It is this very parallel dimension that you are revisiting now mentally. Ulema enter that dimension physically and mentally. But because you are still a student, you will only revisit that dimension mentally and holographically.
- **48**-You are now revisiting and re-living past times in another dimension. Even though, you are experiencing this extraordinary phenomenon and interacting with re-structured events, you will not be able to bring to your physical world the changes you have made in the parallel dimension, because you cannot transport them to a different dimension. They exist only in that holographic reality, which never ceased to exist.
- **49**-However, you will be able to bring back with you the effects and positive results that such changes have created by altering the past. Those results created in a different dimension will materialize physically in your real life, and you will be able to sense and recognize their effects in a very realistic manner, in all things closely related to their nature.
- **50**-In fact, what you have done is changing or totally eliminating the negative and destructive effects of things and decisions you have made in your past. And this is what really counts. Because you have substituted bad results with good results.
- **51**-Those results will continue in your current life when you return to your physical world, and will prevent future bad incidents and bad things from occurring again in your current life. Worth mentioning here that the kind of bad experiences you had in your past will never happen again to you. You have blocked them for good.

III. How real is the holographic/parallel dimension you are visiting?
Is it fantasy?
Is it hallucination?
Is it day-dreaming?
Is it wishful thinking?
None of the above. It is as real as your physical world. Everything you see and feel in this materialistic/physical world exists also (Identically) in other worlds. But their properties are different.

Nevertheless, they look exactly identical. Even the smallest details are preserved. So what you will be seeing during your trip is real. What happened in the past still exists in other dimension, and follows rules that our mind cannot understand.

That incomprehensible and wonderful dimension does not only contain events, peoples, and things from your past, but also whatever is currently occurring on Earth. In other words, you as a human being living on this planet, and working as a teacher or as a physician, you are simultaneously living and working as a teacher or as a physician in a parallel dimension.

You may not have the same name or nationality, but your essence, persona and psycho-somatic characteristics are the same. In fact, you live simultaneously in many and different worlds. Perhaps, and as we speak, there is another person who looks exactly like you in a different dimension is trying to visit you here on Earth, or perhaps, as we speak he/she is so curious to know if you look like him/her, what are you doing, are you aware of him/her in other dimension?

Highly advanced copies of you are fully aware of your characteristics and existence on this planet.

If he/she is older than you in linear time, then he/she is from your past. If he/she is younger than you, than he/she is from your future. But in all cases and in all scenarios, you are one and the same person living at the same time in different places, each place functioning according to different laws of physics, or lack of laws of physics.So, when you revisit your past in a different dimension, prepare yourself to meet another copy of you, or simply YOU in a different non-linear space/time existence.Many adepts who have visited that dimension did not want to return back. In fact, some Talmiz-Ulema remained there. But as you know by now, their copies remained on Earth.

Final words on your visit to the parallel world you have visited using the Dudurisar technique. While you are there, you will not have terrestrial sensorial properties, because your physical body is still in your room, where you began, however, a sort of a spatial memory you will pick up on your way to the parallel dimension, and will substitute for physical feelings and impressions. This spatial memory you had in you all your life, but you never knew you had it, because you have not entered another non-linear dimension before now. It will come handy, and will help you to remember or identify things.

While you are there, you will have all the opportunities of :
a-Meeting many of your past friends (Dead or alive, so to speak in terrestrial terms), perhaps loved pets you lost;
b-Making new friends, and this could sadden you knowing that you will be returning home;
c-Watching the re-happening of future events, as if you were watching a film (you saw before) backward;
d-Learning from new experiences and results you can use when you return home.
The time will come, when you become capable of realizing and understanding the many lives and existences you have now, had, or will acquire in the future.

IV. Some of the benefits:
- **1**-Your mind (On its own) knows when it is time to return home.
- **2**-Some people have reported the sadness they felt when they realized that they had to return home. Many did not want to return home, because they have discovered a better place, made new friends, and lived a life –even though extremely short – free of worries and troubles. This could happen to you. But your brain is stronger than your wishes, and you will return safe.
- **3**-On you way back, your mind will reassure you that everything is going to be just fine, and that you have succeeded in removing some stains, copies of disturbing events from your past, and above all the effects of past actions and events that have handicapped you in this life.
- **4**-Your mind will retain all the knowledge you have acquired, and will guide you more efficiently in all your future decisions.
- **5**-Numerous people have reported that their journey was beneficial, because for some incomprehensible reasons, many of the problems and difficulties they have faced before evaporated in thin air.

IV. Closing the Technique:
- **1**-For this particular technique, you mind will close the technique by itself.
- **2**-You will wake up, and you will realize that you were somewhere else, far, far away from your room.

- **3**-You open your eyes, and all of a sudden, a delightful and reassuring sensation will invade your being. You feel energized, more confident, and in a total peace with yourself.

Harabah: Destruction; ruins. From Harabah, derived the Assyrian adjective Haraba "Kharaba, which means ruins; wasteland.
Kharaab in Arabic, which means destruction, and Kharbah, "Kherbeh", which means rubbles.Kherba in Hebrew, which means ruins.
Harav "Karav" in contemporary Hebrew, which means rubble; ruins. Harabu "Kharaabu in Akkadian, which means devastation. Harbatu "Kharbatu" in Akkadian, which means ruins; deserted area. Krb in Ugaritic/Phoenician, which means ruins. Kherba in proto-Hebrew, which means a wasteland; deserted land.

Haridu "Haridu-ilmu": Interpretation of messages sent to the Conduit in an Anunnaki's or a human's brain cell. Also, it implies to missing or misinterpreting a message by the Conduit.
On this Ulema Rabbi Mordechai said:
- "First of all, you have to remember that your mind (Your brain) has nothing to do with your Conduit. Even though, your brain is functioning wonderfully and you are doing great things in your life, not all the cells in your brain have been used.
- There are so many regions in our brain that have not been explored yet by science.
- In those many unexplored regions of the brain, are so many cells yet to be discovered, located and localized. And above all, we need to learn how they function.
- In that mysterious undiscovered region of the brain, the Conduit exists. It could be in the right or left side of your brain, or just adjacent to line dividing the two parts.
- In the Conduit, there are so many cells, each one with a very defined and particular extraordinary faculty/power, that needs to be activated.
- For instance, one cell triggers the faculty of reading others' thoughts, another cell (Or cells) is responsible for the faculty of teleportation, so on.

- If those cells are not activated, you will not be able to do all those wonderful things.
- So, you have to consider the Conduit as a bank where so many cells are deposited. And there are hundreds of thousands of cells deposited in the Conduit.
- Each cell has a precise function and an invisible location.
- This means that the Conduit can do so many things, if cells are activated. It would be impossible in one lifespan to develop and activate all the cells.
- Three or four fully activated cells is more than enough. With four activated cells you can do four great miracles by earth's standards.
- But for the cell to produce this extraordinary power, the cell must be able first to understand what you want to do.
- For instance, you cannot tell or command your cell "go ahead and make me fly or let me learn a new language in one hour."
- You should first learn how to send your command to your cell. There is a technique for this.
- Your Ulema teacher knows how to put you on the right track.
- Let's assume you have sent a message (A thought, a wish) to your cell. What's next? Well, the message enters your Conduit. Your conduit acting as a supervisor, and as the main receiver reads your message and directs your message to the appropriate cell.
- Your Conduit knows which cell is activated and designed to comply with your request.
- Instantly, the cell receives the Conduit transfer (Meaning your message.)
- Then what? The cell reads your message.
- If your message was sent correctly, then the cell will accept it and give it a code. So, if in the future you ask again your Conduit to do the same thing you have asked in the past, the cell will execute your request in a fraction of a second.
- In other words, each request is coded, and stored in your cell.
- Only coded messages are stored in your Conduit.

- How would you know if you have or have not sent a message correctly to your Conduit? You will know right away. It is very simple. If you have not been trained, you wouldn't know where and how to start in the first place.
- This is the reason why your Conduit did not catch your message(s). You asked "Does this mean that my Conduit is not receiving clear messages from me? And the answer is yes! Your Conduit received something, a thought, a feeling, a wish, a request, call it whatever you want, but your message was not clear to your Conduit, because you did not send your message according to the rules.
- What are these rules?
- They are explained below. But continue to read this first.
- And then you asked: "And how can I send clear messages my Conduit can catch and understand?
- You have to use the "Transmission of Mind" technique. Practice this technique before you send messages to your Conduit. For example, in the past, the SOS (Morse Code) was used by ships, planes, military troops and others. The person who has sent the message (Morse) knew the Code; he/she knew how to tap it. Each word had a code...one dot, two dots, three dots, one dash, two dashes, three dashes, one dit, two dits, one space, two spaces, three spaces, etc. There is a sequence of pulses and marks. And the person who received the message knew how what these dots, and dashes meant. This is how and why he/she was able to read the message or decipher it, if it was a secret message. Your Conduit works exactly in the same way.
- Your Ulema teacher will tell you exactly what dots and sequences to use.
- If your Conduit is hundred percent awake, meaning Open (After training completion), the Conduit will immediately interpret/translate and understand your dots, dashes and sequences.
- Consider those dashes and dots a "Password", a log-in information, a key to open the contact with your Conduit, just like the password you use to open your computer or have access to some websites.
- In the Book of Ramadosh, you will find several passages referring to the brain waves and mind frequencies, and

- some techniques used to direct thoughts and mind energies.
- Your Conduit has its own mode. As long as your Conduit is not activated, it remains free of your control. Once your Conduit is activated, you become the stimulus and the manager of your Conduit.
- The Conduit works partially when it is not activated.
- And partially means reacting by not acting.
- The Conduit functions all the time regardless of your state of awareness, enlightenment or readiness. But it will not give you data and information.
- Everything the Conduit finds is instantly deposited in its compound.
- You will not find what's in there, until the Conduit is fully activated.
- Consider it for now as a depot of knowledge; a sort of a personal bank account where your daily balance is constantly increasing, however, you are not allowed to have access to your bank account.
- So, nothing is lost.
- Your Conduit collects and stores information all the time, and from various sources, times, and spheres.
- (Source: Book of Ramadosh by Maximillien de Lafayette.)

Harima: Title given to a female Anunnaki who is in charge of the education of Anunnaki children. The education and orientation programs of young Anunnaki continue till 71, which is the official maturity age of all Anunnaki in Nibiru. The first 25 years of studying and learning period is usually under the direct control of an Anunnaki female called Harima.

Harimu: Name or title given to the head of an Anunnaki's family cell on other planets.

Harranur-urdi: A term applied to retrieving data and codes displayed on a cosmic monitor called "Miraya". On this subject, Ulema Ben Zvi said (Excerpts from his Kira'at, in the Book of Ramadosh):
- "On Ashtari, the Planet of the Anunnaki (Ne.Be.Ru or Nibiru, to others) each Anunnaki (male and female, young and adult) has a direct access to the Falak Kitbah

(Akashic Records) through the Akashic Libraries (Called Shama Kitbah), which are located in every community in Ashtari.
- The libraries (Called Makatba and/or Mat-Kaba) are constructed from materials such as chiselled opaque glass (Called Mir-A't), a substance similar to fibreglass (Called Sha-riit), and a multitude of fibre-plastic-like materials (Called Fisal and Hiraa-Ti); they convey the appearance of ultra-modern, futuristic architecture (By humans' standards), and techno- industrialized edifices.
- One enters the libraries through an immense hall (Called Isti-bal), seven hundred to one thousand meters in length, by five hundred meters in width.
- The Isti-bal is empty of any furniture, and is lit by huge oval windows that are placed near the top of the ceiling.
- The windows (Called Shi-bak) were designed in such a way that the shafts of light that enter through their circular compartments are redirected and projected like solid white laser beams. The effect is spectacular.
- At night, quasi identical effect is produced by the projection of concentrated light beams coming from hidden sources of lights located behind (More precisely, inside the frames' structure) the frames (Called Mra) of the windows.
- The frames serve as an energy depot.
- The energy is transformed into sources of light.
- The visual effect is stunning. Enormously large and animated metallic billboards (Called Layiha, pronounced La-ee-haa) are affixed on walls in a parallel alignment, and on the floor, in front of each billboard, there are hundreds of symmetrically rectangular pads (Called Mirkaan).
- When visitors enter the library's main hall (Situated just at the front entrance), they approach the billboards, and stand each on a Mirkaan.
- The pad serves as a scanner and a transportation device, because it has the capacity to read the minds of the visitors, learn what they are searching for, and as soon as it does so, it begins to move, and slides right through the central billboard (Called Kama La-yiha), which is not

really solid but is made from blocks of energy, carrying the visitor with it.
- Behind the billboard is the main reading room (Called Kama Kira'at) of the Akashic Library.
- Under the belly of the pad, there are two separate compartments designed to register what the visitor is looking for, and to direct the visitor to his/her destination; usually, it is a reference section where books in form of cones are located on magnetized shelves.
- The Anunnaki's Akashic Library is not a traditional library at all, for it contains no physical books per se, even though, there are plenty of conic publications (Books manufactured as magnetic cylinders and cones.) Instead of searching for books on shelves, as we do on Earth, the visitors find themselves in the presence of an immense white-light blue screen, made of materials unknown to us.
- The screen is hard to describe; it can be compared to a grid (Called Kadari), with a multitude of matrices and vortices of data.
- The visitors communicate with the screen via their Conduit. The screen registers their thoughts and right away finds/records the information the visitors seek.
- All the visitors have to do is stand still for less than two seconds (In terrestrial terms) in front of the screen, and the data will be displayed in an animated format.
- The data (Information) is given in codes which are easily understood by the visitors.
- The codes are usually divided into sequences; each sequence reflects an aspect of the information.
- For example, if you want to know what happened in Alexandria or Phoenicia 3000 B.C., all what you need to do, is to think about either Alexandria or Phoenicia, and one grid will appear, waiting for your command to open it up.
- From this precise moment, the visitor's Conduit and the Screen are communicating in the most direct fashion.
- The grid opens up and displays three files in sequence.
- The nearest description of these files would be plasmic-digital, for the lack of the proper word; each file will

contain everything that had happened pertaining to that particular date or era in Alexandria or Phoenicia.

Functions of the Conduit, Miraya and retrieving data:
- **1**-The Conduit will sort out all the available information and references (Photos, holographic projections, sounds) available on the subject. On Ashtari, everything is stored in codes.)
- **2**-The Conduit selects and indexes the particular data for the part of the information the visitor is most interested in.
- **3**-Then, the selected information (Complete data in sound and images) is instantly transferred to the cells of the visitor's brain.
- **4**-Because Anunnaki are connected to each other and to their community via the Conduit, the data recently absorbed is sent to others who share similar interest.
- **5**-This is extremely beneficial, because if the data received from the screen is difficult to understand, other members of the Anunnaki community, will automatically transmit, the explanation needed. This is quite similar to an online technical support on earth, but it is much more efficient since it functions brain-to-brain.
- **6**-Each Anunnaki community has the same kind of center for these Akashic files.
- **7**-The complexity of the centers, though, is not the same. Some of the Akashic Libraries include more perplexing and complicated instruments and tools, which are not readily available to other communities.
- **8**-These tools include the Monitor, which is also called Mirror, or Miraya. Each Miraya is under the direct control of a Sinhar, who serves as custodian and guardian.
- **9**-The screens can expand according to the number of codes that the Anunnaki researcher is using. Seven to ten codes are normal.
- **10**-If a larger number of codes are opened, the screen is fragmented into seven different screens. An amazing phenomenon occurs at this moment – time and space mingle together and become unified into one great

continuum. This enables the researcher to grasp all the information in a fraction of a second.
- **11**-An added convenient aspect of the Akashic files is the ability of the researchers to access them in the complete privacy of their homes or offices, since part the files can be teleported there.
- **12**-But since the private screen is not as complicated as the central one in the Library, no multiple screen will open up, only the original one. Yes, Anunnaki do live in homes, and contribute to their societies as we do on Earth.
- **13**-It is important to understand that the data received is not merely visual. There is much more to it than that.
- **14**-By the right side of the screen, where the global data is displayed in files, there are metallic compartments, as thin as parchment paper, which serve as a cosmic audio antennae.
- **15**-These compartments search for, and bring back, any sound that occurred in history, in any era, in any country, and of any magnitude of importance.
- **16**-The compilation includes all sorts of sounds, and voices of people, entities, various civilizations, and living organisms (And life-forms from the past and the future) in the entire cosmos.
- **17**-According to the Anunnaki, every single sound or voice is never lost in the universe. Of course, it may not traverse certain boundaries. For humans, if a sound was produced on earth, such a boundary is the perimeter of the solar system.
- **18**-Each of these antennae-compartments will probe different galaxies and star systems, listening, recording, retrieving, and playing back sounds, voices, and all sorts of frequencies.
- **19**-A combined asset of the visual and audio systems is the ability to learn languages that is afforded by the Akashic Library. This applies to any language – past, present or future, and from any part of the universe.
- **20**-The researcher can call up a shining globe of light that will swirl on the screen with enormous speed. As it rotates, the effect blends with an audio transmission that

comes from the metallic compartments, and in an instant, any language will sink into the brain cells.

- **21**-On the left side of the main screen, there are several conic compartments that bring still images pertaining to certain important past events. This display informs the researcher that these particular events cannot be altered. In other words, the Anunnaki cannot go back into the past and change it.
- **22**-The Anunnaki are forbidden to change or alter the events, or even just parts or segments of the past events projected on the main screen that came from the conic compartment, if the data (Images; sounds) represents events created by the Anunnaki leaders.
- **23**-This restriction (Altering, changing or erasing past events) which is applicable everywhere on Ashtari functions as a security device.

  For example, a young Anunnaki cannot visit earth sixty five thousand years ago, recreate and enter the genetic laboratory of the Anunnaki in Sumer, Arwad, Ugarit or Phoenicia, and change the DNA and the genetic formula of a human race, especially when the DNA sequence was originally created by an Anunnaki Sinhar (Leader). In other words, a young Anunnaki is not allowed to alter the Akashic Records that contain the primordial events. Alteration such as recreating a new human race in past time will never happen.

  24-However, an Anunnaki leader such as Sinhar Baal Shamroot, Inanna, Ellil or Enki can go back in time and space, and change events, but cannot bring to Earth new human species created according to a new formula that contradicts or reverses the primordial prototypes created 100,000 years ago.
- **25**-However, an Anunnaki leader such as Sinhar Baal Shamroot, Inanna or Enki can go back in time and change events, but cannot bring to Earth new human species created according to their new formula and based on the primordial prototypes created 100,000 years ago. Nevertheless, they can transpose their new creation or event alteration, and transport them to another dimension, parallel to the original dimension where the event occurred.

- **26**-This safeguard means that Sinhar Inanna cannot recreate a new race on our earth by the device of sending the current living humans back in time, remolding us, and then bringing us back to the twenty first century as a new species.
- **27**-This is not allowed by the Anunnaki High Council. All she can do is recreate her own experiment in another dimension.
- **28**-Worth mentioning here, that alternation of the fabric of time and space is rigidly and constantly monitored by the Anunnaki High Council via their Miraya; the cosmic mirror and monitor of all living-forms, past, present, and future.
- **29**-The Miraya is a terrific and mind-bending tool. The Anunnaki use it to revisit time and what is beyong time, space, meta-space, and para-space, as well as creating new cosmic calendar.
- **30**-More options are available for research, and one of them is a sort of browsing. Inside the screen, there is a slit where the mind of the Anunnaki can enter as a beam. This will open the Ba'abs, or Stargates, to other worlds that the researcher is not even aware of; they appear randomly as part of the discovery or exploration.
- **31**-In each slit, there is another Akashic file that belongs to another civilization.
- **32**-Sometimes, these civilizations are more advanced than the Anunnaki themselves, where the researcher can retrieve important information. It is like going back in the future, because everything present, or to occur in the future, has already occurred in a distant past (Timetable) and needed the right time to surface and appear before the current living Anunnaki.
- **33**-There is also the aspect of simply having fun, some of which is not so ethical. Sometimes an Anunnaki will go back in time, let's say 400 C.E., choose a famous historical figure, and at the same time bring over another important person, one thousand years older, simply to see how they would interact.
- **34**-They can easily deceive these personages, since every Anunnaki is an adept at shape changing. Or they can

- transpose people, move them in time, and see how they will react to the new environment.
- **35**-These games are strictly forbidden, but some low class Anunnaki occasionally try it as a game. Sometimes they interfere with our daily affairs, and temporary loss of memory may be a result of that.
- **36**-Worth reminding the readers, that the Anunnaki no longer interfere in human affairs. They have left planet Earth for good, but they are coming back in 2022.
- **37**-The Miraya is constantly used by the Anunnaki on Ashatari. In addition to its function as a cosmic calendar, the Miraya serves as a galactic monitor. Watching and monitoring other extraterrestrial civilizations are two of the major concerns of the Anunnaki.

For the past 10 years (In our terrestrial time), the Anunnaki have been following very closely what was/is happening on other planets and stars, and particularly the experiments of the Greys. Sinhar Ambar Anati known to us as Victoria told us a lot about this. In the Book "Anunnaki Ultimatum: End of Time" (Co-authored by Ilil Arbel and Maximillien de Lafayette), Ambar Anati described at length some of the horrible genetic experiments of the Greys as caught on the Miraya.

Note: Ambar Anati is talking to Sinhar Inannaschamra, her mother-in-law in Ashatari, about the malicious intentions of the Greys. Inannaschamra told her that the Greys constantly conduct genetic experiments on humans, and that the Anunnaki kept on watching their atrocious experiments by using the Miraya (Monitor).
Here are some excerpts from the book: "What do they want from us on earth?""There are a few things that they want. First, they want eggs from human women and use them to create hybrids. Let's take a look at this monitor, and I'll show you how they do that. But Victoria, steel yourself. This is pretty horrible, even though I have seen even worse. You will also be able to hear, it is like a television."The monitor blinked and buzzed, and a small white dot appeared on the screen. It enlarged itself, moved back and forth, and settled into a window-like view of a huge room, but the view was still rather fuzzy. I heard horrendous screams and froze in my seat, these were sounds I have never heard

before. After a few minutes the view cleared and I saw what seemed to be a hospital room, but it was rounded, not square. Only part of it was revealed, as it was elongated and the far edge was not visible. The walls on the side were moving back and forth, like some kind of a balloon that was being inflated and deflated periodically, with a motion that made me dizzy; they seemed sticky, even gooey. The room was full of operation tables, of which I could see perhaps forty or fifty, on which were stretched human beings, each attached to the table and unable to move, but obviously not sedated, since they were screaming or moaning.

Everyone was attached to long tubes, into which blood was pouring in huge quantities. I noticed that some of the blood was turning into a filthy green color, like rotting vegetation. At the time I could not understand what that was, but later that day I found out. This blood was converted to a suitable type for some of the aliens that paid the Grays to collect it, and it was not useful in its raw condition.

People who operated these experiments were small and gray, and they had big bug eyes and pointy faces without any expression. I thought they looked more like insects than like a humanoid species. They wore no clothes, and their skin was shiny and moist, like that of an amphibian on earth. It visibly exuded beads of moisture which they did not bother to wipe away. Each operating table had complicated machinery that was poised right on top of the person who was strapped to it.

On some of the tables, the machinery was lowered so that needles could be extracted from them automatically, and the needles reached every part of the human bodies, faces, eyes, ears, genitals, stomach. The people screamed as they saw the needles approaching them, some of them fainted. Many of the people were already dead, I could swear to that. Others were still alive but barely so, and some had arms and legs amputated from their bodies.

It was clear that once the experiment was over, every single person there will die. I don't know how I could continue to look, but somehow I managed. I looked at the ceiling of this slaughter house and saw meat hooks, on which arms and legs and even heads were hanging, like a butcher's warehouse.

On the side of the tables were large glass tanks where some organs were placed, possibly hearts, livers, or lungs, all preserved in liquids.

The workers seemed to be doing their job dispassionately and without any feelings, moving around like ants and making buzzing sounds at each other as they conversed.

They were entirely business-like and devoid of emotion. At least, their huge bug eyes did not convey any emotion to me, neither did their expressionless faces. I watched until I could no longer tolerate it, and finally covered my eyes and cried out, "Why don't you stop it? Why don't you interfere?" Sinhar Inannaschamra turned the monitor off. "This event is a record from decades ago, Victoria. It is not happening now as we look. And even though often we do interfere, we cannot police the entire universe or even the entire earth.

They know how to hide from us. And you must understand, that often the victims cooperate with their abductors."

"Why would they?"

"Basically, through mind control. The Grays have many ways to convince the victims. The Grays can enter the human mind quite easily, and they find what the abductees are feeling and thinking about various subjects. Then, they can either threaten them by various means, or persuade them by a promise of reward."

"Reward? What can they possibly offer?"

"Well, you see, they show the victims images through a monitor, just like this one. They tell them that they can send them through a gate, which is controlled by the monitor, to any number of universes, both physical and non-physical.

That is where the reward come in.

For example, if the abductees had originally reacted well to images of Mary or Jesus, the Grays can promise them the joy of the non-physical dimensions.

They show them images of a place where Mary and Jesus reside, where all the saints or favorite prophets live, and even the abode of God. They promise the abductees that if they cooperate, they could live in this non-physical universe in perpetual happiness with their deities. Many fall for that."

"And if they resist?"

"Then they show them the non-physical alternative, which is Hell. Would you like to see some of it?"

"You can show me Hell?" I asked, amazed.

"No, there is no such thing as Hell... it's a myth that religions often exploited. But I can show you what the Grays show the abductees, pretending it is hell; they are quite devious, you know.

You see, some creatures live in different dimensions, where our laws do not apply.

Sometimes, they escape to other dimensions.

These beings have no substance in their new dimensions, and they need some kind of bodies to function. At the same time, the Grays can tap into numerous universes, because they can control their own molecules to make them move and navigate through any dimension. Well, a cosmic trade had been developed.

The Greys supply the substance taken from human abductees, and from the blood of cattle. You must have heard of cattle mutilation, where carcasses of cows are found in the fields, entirely drained of blood? The Grays do it for their customers."

"How do these creatures pay the Grays?"

"By various services, and once they get their substance, they are incredibly powerful in a physical sense.

The old tales of genies who can lift buildings and fly with them through the air were based on these demons; the Grays often have a use for such services.

But let me show you a few of these creatures. Of course, you can only see them when they have already acquired some substance from the Grays."

The monitor hummed again as Sinhar Inannaschamra turned it on. The white dot expanded into its window, which now, for some reason, was larger and took over the entire screen.

All I could see was white fog with swirls floating through it. Sometimes the fog changed from white to gray, then to white again.

I started hearing moans. Not screams, nothing that suggested the kind of physical pain I saw before, but perhaps just as horrible, since they voices where those of hopelessness, despair, and emotional anguish. Every so often I heard a sound that suggested a banshee's wail, or keen, as described in Irish folklore.

"It will take a while for someone to show up," said Sinhar Inannaschamra. "Most of them have no substance, and therefore they are invisible. Others have a shadowy substance.

Then, there are the others...but you will see in a minute. Once they notice they are being watched, they will flock to the area, since they are desperate to get out.

Incidentally, it was never made quite clear to me how they produce sounds without bodies, we are still trying to find out what the mechanics are, but it's not easy, because we would rather not go there in person." "They sound horribly sad," I said.

"This is what makes it so hell-like.
In many cultures, Hell never had any fire and brimstone and tormenting devils, but rather, it was a place of acute loneliness, lack of substance, and alienation from anything that could sustain the individual from a spiritual point of view. Think of the Greek Hades, or the ancient Hebrew Sheol, before the Jews made their Hell more like the Christian one. Look, here comes the first creature. Poor thing, he is a shadow."
I saw a vaguely humanoid shape in deep gray. It seemed to have arms, which it waved in our direction. It seemed fully aware of the monitor.
Then another shadow, then another, all shoving each other and waving desperately at the monitor.
Then something more substantial came into view, and I jumped back as if it could reach me. It seemed to be a severed arm. Cautiously, I came back, and then saw that the arm was attached to a shadow body. I looked at Sinhar Inannaschamra, speechless, and she said, "Yes, here you see one that managed to receive an arm. It wants to complete its body, of course, so that it can get out of this dimension and serve the Grays, but the Grays keep them waiting until they want them." (Source: Anunnaki Ultimatum: End of Time.) In summary, the Miraya has multiple functions, and remains one of the most important inventions of the Anunnaki.

Hatani: A shield; a protection.
Usually referred to a plasmic halo surrounding the physical body of an Anunnaki-Ulema. From Hatani, derived the Akkadian verb Hatanu, which means to shelter. Hatani is closely related to the Anunnaki-Ulema Baridu technique.
Baridu is the Anunnaki-Ulema term for the act of zooming into an astral body or a Double.

I. The Concept:
Anunnaki-Ulema explained Bari-du as follows:
Verbatim; Excerpts from a Kira'at:

- The initiated and enlightened ones can zoom into their other bodies, and acquire Anunnaki supernatural faculties.
- I have used the words supernatural faculties instead of supernatural powers, because the enlightened and initiated ones are peaceful, and do not use physical

- power, brutal force or any aggressive means to reach their objectives.
- The use of violence against humans and animals, even aggressive thoughts and harmful intentions annihilate all chances to acquire Anunnaki's extraordinary faculties.
- Your Double can easily read your thoughts.
- If your thoughts are malicious, your Double will prevent you from zooming yourself into its ethereal molecules.
- Therefore, you have to control your temper, remain calm, and show serenity in your thoughts, intentions and actions.
- You Double is delicate, even though it can accomplish the toughest missions and penetrate the thickest barriers.
- Any indication of violence or ill intention triggers a pulse that blocks your passage to the ethereal sphere of your Double.
- Once you enter your Double, you will be able to use it in so many beautiful and effective ways as:
  1- A protective shield against danger,
  2- An effective apparatus to protect yourself in hostile and dangerous situations,
  3- A tool to develop your abilities to learn many languages, and enhance your artistic creativity,
  4- A stimulus to increase the capacity of your memory,
  5- Instrument to heal wounds and internal injuries. No, you will not become a surgeon, but you will be able to stop internal bleeding, and eliminate pain,
  6- A vehicle to visit distant places and even enter restricted areas for good causes. The possibilities are endless.
- Once you are in a perfect harmony with your Double, and your physical organism is elevated to a higher vibrational level through your union with your Double, you will be able to walk through solid substances such as walls, sheets of glasses and metal.
- You become effective in controlling metal and de-fragmenting molecules of any substance. This will allow you to transmute, change and alter the properties of any object known to mankind.

- But if you use these supernatural faculties to hurt others, or for personal and selfish gain, you will loose them for good, and you will be accountable for such malicious use in the other dimension.
- And this could delay your entrance through the Ba'ab.

II. The Hatani Protection Shield:
Excerpts from Ulema Kira'at and the Book on Ramadosh; Ulema Govinda said about the protection against threats and dangers (Verbatim):

- Only those who have learned and developed esoteric Ulema techniques can use their Double as a shield.
- However, a novice or a sincere student who is searching for the ultimate paranormal truth, and who has revealed a high standard of spirituality and goodness will be able to use the Double, once he/she has completed the Ulema studies.
- At a novice stage, the Double is alive and well, and is fully aware of your existence, but as a novice, you are not aware of its existence, because you have not established a rapport with your Double.
- Once, a rapport has been established with your Double, your Conduit will throw an invisible protective shield around you.
- But was is a shield? Is it a physical barrage? A protective tool or a device similar to the fibreglass or a metal shield police use in riots? The answer is no.
- The shield functions in so many different ways your brain cannot comprehend. However, I will try to explain to you one of the protective measures a shield uses in threatening situations.
- The Ulema after years of study and practice, and following the instructions of the Book of Rama-Dosh, became capable of creating a sphere (Or zone) around them that resembles a halo. Some call this halo a "Bubble".
- The halo surrounds their physical body.
- In the halo, exist molecules and particles charged with high atomic and sub-atomic density (No, not nuclear devices!), i.e. energy.

- This energy is denser in its composition than any of the molecules and particles that physically create and constitute any physical action or movement against an Ulema's body.
- Ulema's energy changes constantly and transmutes itself into higher or lower molecules and particles density, according to their surroundings and needs.
- Because of the Ulema's denser atomic substance, nothing can penetrate the halo surrounding them.
- You have to remember, that everything in the universe is composed from molecules and particles. For instance, if you throw a punch at an Ulema, you put in your punch a certain amount of energy and physical effort. The energy and the physical effort are composed from molecules.
- These molecules are denser in their composition than the molecules floating around the Ulema, and thus cannot penetrate their halo and reach their bodies.
- This is why people can't go through walls. Our bodies molecules need "to shrink" and "transmute" themselves into thinner vibrations, to allow us to go through walls.
- The non-physical entity (Double) you have mentioned is not totally non-physical. It changes. It materializes and dematerializes.
- The Double can project itself as a physical entity. And it takes on multiple appearances ranging from holographic to multi-dimensional presences.
- However, the materialistic apparition does not last very long in a three-dimensional sphere, because its bio-etco-plasma energy is consumed rapidly.

In other words, and simply put,
- **a**-Your Double is extremely intelligent and alert, and it senses things around you. Things currently happening and those en route.
- **b**-Your Double knows right away if what is coming at you is safe or dangerous.
- **c**-If the Double detects a threatening situation, it sends an alert to your Conduit.
- **d**-Your Conduit receives the message from your Double. (Note: Sometimes, it is simultaneously, and/or your Open Conduit understands the situation on its own, without the help of your Double.)

- **e**-Your Conduit acts on its own and guides you instantly to a safer position. Call it whatever you want, instinct, an inner feeling, etc...it does not matter what you call it.
- **f**-At the same time, your Conduit emits vibes aimed at the source of the threat to bock it.
- **g**-At this very moment, your Open Conduit and your Double act in unison.
- **h**-In a fraction of a second, the attacker or the negative vibes aimed at you is diverted. Nothing can penetrate the halo around you. If you practice and master the Baridu technique, you will be able to block any threat." (Source: The Book of Ramadosh: 13 Anunnaki-Ulema Mind Powers Techniques to live Longer, Happier, Healthier, Wealthier, authored by Maximillien de Lafayette.)

Hatori "Hatori-shabah": The metamorphosis and changes of the face's skin. It refers to various phases of youth brought to a person via the practice of the Anunnaki-Ulema Daemat-Afnah technique. It is closely related to Daemat-Afnah, which is a term for longevity, and halting the process of aging.
It is composed of two words:
a-Daemat, which means longevity.
b-Afnah, which means many things, including health, fecundity, and longevity.
According to the Anunnaki-Ulema, we are not programmed to age.
By practicing the Daemat-Afnah, a person will regain his/her youth, and his/her face will look 37 year old. On the subject of Hatori-shabah, Ulema Suleiman Al Bak'r said (Verbatim):

- "You have to practice the technique for at least one full year. You will not see any improvement or any result before 12 months. I will explain to you what is going to happen step by step.
- For the first 6 months. You will not notice any change on your face.
- At the end of the seventh month, you will begin to feel that some of your facial muscles are getting stronger. A strange and a new sensation you have never felt before in your whole life.
- Your face will look cleaner and firmer.

- Some of the wrinkles under your eyes will disappear. In rare instances, they would not.
- Not all the wrinkles will disappear if you stop practicing.
- At the end of 12 months, you will notice that you eyes have gained vitality. They will look sharper.
- Your eyes will glitter with a sign of good health.
- At the end of 12 months, you will notice that your face's skin is healthier, and almost 90% of your wrinkles (large and small) have diminished.
- The dermatologic results have no side effect.
- A certain incomprehensible inner strength will energize your whole body.
- After 13 months of practice, the face you had when you were 37 starts to reappear gradually. You will not believe what you are seeing.
- This change is usually accompanied by sizeable increase in physical dynamism and mental vitality.
- Your face is younger, almost 100%.
- Only your face gets younger, not your neck, body or any other part of your body.
- Your grey hair will stay grey.
- If you are bold, you will stay bold."
- You will keep your new face for a very very long time, as long as you keep a good diet, and you eat well. (Source: The Book of Ramadosh: 13 Anunnaki-Ulema Mind Powers Techniques to live Longer, Happier, Healthier, Wealthier, authored by Maximillien de Lafayette.)

Hattari: The first stage of the afterlife during the 40 day period following death. In that stage, a new life-form develops in the mind of deceased people. A brief description of Hattari was given by a lady (Who apparently was an Ulema) to her beloved son. Here is an excerpt from the "Book of Ramadosh", and from the Book "On the Road to Ultimate Knowledge: The Anunnaki-Ulema Extraterrestrial Tao", co-authored by Dr. Ilil Arbel, and Maximillien de Lafayette:
The scene is between Germain Lumiere, an Ulema from France who has just lost his mother.
It appeared later on, that his mother was an Ulema too, but she has never told him that, for reasons we don't know.

Two days after she passed away in Paris, his mother appeared to him during her funeral, as she has promised him.

The young Ulema asked his mother lots of question about the after-life, and what is she doing there. Herewith, a brief excerpt from their conversation:

Location: Cemetery of Père-Lachaise, Paris, France.

Time: In the afternoon, during the funeral of Countess....mother of Germain.

Personages:

1-The deceased mother appears as a spirit and talks to her son Germain, while her physical body is in the coffin.

2-Germain in tears talking to his dead mother for the last time.

3-Sylvie: She is Germain's sister.

Excerpt below: Germain is telling us what they talked about at the funeral.

"I returned to Mama, who was looking sadly at Sylvie. It's really too bad I can't talk to her," Mama said to me, "but some day, of course, she will know, like everyone else. Ah, well, let's go to the more secluded areas. We don't want people to think you are talking to yourself."

We wandered around the cemetery. Père-Lachaise, is one of the most beautiful cemeteries in the world, full of trees, impressive statues, and old tombstones. Shady lanes provided privacy, and we could talk freely.

"So tell me about your experience in the Afterlife, Mama," I said.

"I have not been there very long, you know, but time and space play a different role there, and also, my training allows me to know what it is really like and what will happen next," said Mama. "You will also know, when the time comes."

"Doesn't everyone know?"

"No, many of the dead don't realize that they are dead. They don't seem to see the border between life and afterlife. These people can be very anxious. They sometimes try to get back to Earth, meet their loved ones, and they are very upset when the living cannot see them."

"So what happens to them?"

"The guides, spirits of higher dimensions, help them realize that they are dead. Sometimes, if persons have a real need to go back to Earth to accomplish something, the guides are saddened by their pain, and allow them to go back, manifest, and complete their task. Once they do that, they can come back, much happier and calmer. It only happens once, of course, but after that they are ready to adjust to the afterlife."

"What is it like, over there? Were you scared when you passed on?"

"There is nothing frightening about the afterlife," said Mama. "It is very much like earth, but peaceful, much more beautiful, and there is no strife or violence of any kind. To the departed, who have shed their bodies and are occupying a new body, it is as physical as the earth is to the living. Everyone is healthy, there is no disease, no pain, no violence.

There are cities with streets and buildings, gardens and parks, countryside – all seems normal, like a poetic interpretation of life. What you see here is visual projections.

You see millions of real people, coming and going in huge waves. There is much to do, since the place you come to first is no more than a quick stop. You only stay here for twenty to thirty days, some times forty days, and then move on."

"Do they know where they are going?"

"It depends. Most people cannot see what is ahead of them, only what is behind them. But they always move on to a higher phase."

"So naturally they are a bit scared of the unknown."

"Yes, some of them experience anxiety. That is what the twenty to thirty days period is for, deciding what needs and things to be done. And they are helped by the guides, or by people who chose to stay longer in this place."

"So you can stay there longer?"

"Yes, there are various options, of course. One option is to go to the place you have created when you built your "Minzar" and planned a place of rest and happiness. Many people choose to go there for a while – it is up to them how long they would stay there. Time is not really a very important issue where we are. It seems to me that time has stopped. You can stay there forever if you like it very much."

"The place created with the Minzar must be very appealing to most people, I should say," I said. "It's custom made for your own happiness."

"Yes, and the person already has friends, a place to stay, things to do, anything he or she likes best. It's a good option. But eventually, I would say one should try to evolve into the higher dimensions. You don't know what you miss unless you see it."

"When I built the Minzar, Rabbi Mordechai told me that I could not stay in the place I created for too long, since the energy would dissipate and the living body will call me back. But I suppose it's different when one is dead."

"Yes, since this is now part of the depot of knowledge located in your brain, which was created by the Minzar experience. It is your Spatial Memory, my son."

"So you plan to move on after the thirty days?"

"Yes. It is as it should be, and I want to evolve into the higher dimensions. But as I promised, I will come back for you and be your guide when it is your time to follow me. Think about it as a short, though necessary separation, but temporary all the same. What it all comes down to, Germain, is that there is no death. And the afterlife offers so many opportunities for new growth, new knowledge. There is nothing to fear."

"Will you see Papa? Will I see him when I go there?"

"Of course we will. Do not worry and do not mourn me, Germain."

"I will try not to, Mama. I promise."

"Well, my son, I will be leaving now. No need to say goodbye. Rather, au revoir."

I closed my eyes, not wishing to see her leave, and felt something brush my cheek as if she kissed me. When I opened my eyes, there was no sign of her. She was gone. I went home and helped Sylvie attend to the visitors; I have never felt so numb."

Hattoba "Ha-toobah": Term for the process involving the preparation of meeting Djins or Afarit, allegedly, the guardians of the underground room of the Book of Ramadosh, located in the forbidden city of Baalbeck It also refers to the holographic printing process of the Book of Ramadosh. This Anunnaki-Ulema book is not accessible to the general public. And few Ulema of the higher rank are allowed to read it. There are no printed copies of the Book of Ramadosh in Ulema centers and lodges of Les Peres du Triangle (Affiliated with the Ulema). However, there are only two copies available to the Munawareen (Enlightened Ones) and Ulema above the rank of the 18th degree in France, and in the Near East.

Germain Lumiere, a contemporary French Ulema told how he was able to visit the underground city of Baalbeck, and witness the supernatural printing process of the Book of Ramadosh, after a frightening encounter with non-terrestrial entities, who lived underground, right before the long tunnel leading to the room of the Book. His depictions were extraordinary, and a segment of his revelations appeared in the book "On the Road to Ultimate Knowledge", co-authored by Maximillien de Lafayette and Dr. Ilil Arbel. Here are some brief excerpts from this book:

The place: Damascus, Syria.
The scene: Germain Lumiere with his tutor and mentor Ulema Master Li who came to Damascus to visit his student Germain. At that time, Germain Lumiere lived with his mother the countess and his sister Sylvie in Syria, Damascus.
The year: Around 1956 or 1957.
Text: Germain Lumiere telling us the story, in his own words.
"At that time, the Master was visiting us, and as usual, had an incredibly exciting plan for me.
"Have you ever been to Baalbeck?" he asked.
"No, never."
"It's an interesting city, very old. There is a lot of controversy as to who built it, though."
"Isn't there some historical evidence?"
"Plenty, but there are four interpretations.
The Christian Lebanese say it was built by the Phoenicians. The Muslim Lebanese prefer a theory claiming it was built by Djinn and Afrit. Some important occult leaders say it was built by Adam, after he was kicked out of Paradise. Well..."
"And the Ulema, what do they think?" I asked, knowing that this was the theory I would trust.
"The Ulema say it was built by the Anunnaki and the proto-Phoenicians who lived on the island of Arwad and in Tyre. There is a lot of evidence in this direction."
"So will I see the ancient parts?"
"Of course. I would like to take you to a very special part of the city, where the Founding Fathers of the Ulema used to meet thousands of years ago. Unfortunately, we no longer meet there, because it became a tourist attraction and a state-controlled center of music and dance festivals. It will be fun for you, though, to mingle with all these tourists, it's a nice place."
"But surely that is not the reason for going," I said.
"No, it is not. I plan to take you to a secret underground city under Baalbeck, and show you where the Anunnaki landed for

the first time on earth. Very few people know what is going on under the modern city of Baalbeck. The first Anunnaki landing took place before the Deluge, though they came again and again after the Deluge as well."

"Before the Deluge? When was that, exactly?" I asked.

"About 450,000 years ago, perhaps a bit longer. At that time, the Anunnaki created the humans."

"And what about God?" I asked. Even though I was taught much of the Ulema traditions and world view, I never heard about the creation of the human race.

"No one ever heard of God 450,000 years ago. You start to hear about God only around 6,000 years ago," said the Master. I knew enough about the Anunnaki at that time to accept this without much trouble, so I went to find Mama and Sylvie and tell them about the upcoming trip.

The trip from Damascus to Baalbeck could be accomplished in about two hours, at least you could do that if you traveled in a decent car. We took a bit longer to get there, since the car, borrowed from a friend of the Master who was also to drive us there, was an ancient Mercedes that did not use normal gasoline but rather employed *mazut*, or diesel fuel, and made such a racket it was impossible to hear yourself think. To my surprise, I saw a mysterious Sudanese man sitting in the back seat, dressed in ill matching jacket and pants and scowling at us.

At the Master's request, he started to get out of the car to introduce himself. I watched the process in fascination, since he was not doing it quickly like a normal person, but instead was slowly extricating himself in stages, gradually disentangling himself, like a huge snake. I have never seen such a tall man, or anyone as strange. He was about seven feet tall, very thin, and his face did not look quite human to me, but like a giant from outer space.

This bizarre apparition just stood there, looked fierce, and played with a string of amber beads. The Master ignored his uncouth behavior and introduced us.

"This is Taj," he said. "His name means 'Crown.' He is joining us because he has the key to the gate of the secret city underground. He is also able to persuade the Djinn and the Afrit to open certain doors, which is quite a talent." I was not sure if the Master was joking about the Djinn and the Afrit, so I kept quiet, nodded to the Sudanese, and got in the back seat. Taj folded himself back into the car and sat beside me, the Master went into the front seat, and the driver, who seemed to be normal and cheerful, greeted the Master and me in a friendly way. The car started making a noise that was worthy of demons, but I did not care because I was thinking about the real devils, the Djinn and the Afrit.

I leaned forward and asked the Master, "Would I be able to see the Djinn and Afrit?"

"Yes, of course," said the Master casually. "You can even try to talk to them, if you like. The underground city is actually called the City of the Djinn and the Afrit; plenty of devils are there." Since these devils did not seem to frighten the Master, I assumed he knew what he was doing, and sat thinking about what my part could be in this unbelievable adventure. However, I was aware of increasing irritation by what Taj was doing. He constantly played with his amber beads, clicking away on and on. I asked, "Why do you have to click these things all the time?"

Taj seemed annoyed by my question. "Try them yourself," he said curtly, and handed them to me.

I grabbed at them, and instantly, a horrible electric shock went through my entire body, quite painfully, and I cried out and threw the beads on the floor of the car. The Master screamed at Taj, "How dare you? How many times did I tell you never to do that? Give me the beads immediately!"

Taj handed him the beads, meekly enough, and had the grace to look embarrassed. The Master rubbed the beads, seemingly absorbing and removing the energy, and then returned them to me. "You can try them now," he said. "And don't give them back to Taj until I tell you to." Taj said nothing. He seemed unhappy in the car, constantly fidgeting, and could not sit still. Perhaps he was claustrophobic, I thought, and the confined space bothered him. We drove on.

Finally we arrived in Baalbeck. "Where now?" said the driver.

"We are going to the *Athar*, the ruins," said the Master.

"I don't know how to get there," said the driver. "Shall I ask for directions?" He parked the car. There were many people around, some Arabs in traditional garb, some Europeans in every kind of attire and carrying backpacks and cameras. It seemed to be such a normal, cheerful place. I thought of the festivals and the music; how could there be Afrit and Djinn and all sorts of underground labyrinths in a place like that? It was as modern as can be.

"When you are with Taj, you do not ask for directions," said the Sudanese with a superior air. The driver shrugged, not quite convinced.

Taj winked at me and stared at the driver's neck, concentrating. The driver suddenly started to beat his own neck, complaining how much he hated mosquitoes. I was certain there were no mosquitoes in the car, and I was sure that Taj created the imaginary insects that were tormenting the driver. The driver's neck became really red.

"Taj, stop this nonsense immediately!" said the Master severely. Apparently, Taj could send certain energy rays that had the capacity of annoying people. Taj stopped, gave the driver the necessary directions, and we went to the Athar.

"First, let's go to the world biggest stone," said Taj. We drove further, and as we turned a road toward the Temple of Jupiter, I was shocked by the sight that met my eyes. It was a huge gray slab, partially buried in the sand, perfectly cut and smooth. It was unquestionably man made, not a natural formation, a short distance from the Temple. How in the world could such a stone get there? Who could have carried it? This stone was so immense that the stones of the Egyptian pyramids would be infinitely small, completely dwarfed, if put next to it. The Stonehenge monoliths would be insignificant if they were placed next to it. In addition, it was immensely old, and even modern equipment could hardly cope with such a giant, let alone ancient technology.

"How big is this stone?" I asked, truly awed by the sight.

"Seventeen hundred tons," said the Master.

"It is hand made, isn't it?" I said. "It is too straight to be natural. It simply can't be natural. And yet, how could it get here, if it is artificial? It just can't!"

Taj grinned and said, "Hand made, yes, but not by human hands."

I was beginning to get the idea. "Then who made it?" I asked.

"It was part of the landing area used by the Anunnaki," said the Master. "There are six stones like it. Only the Anunnaki could move such a slab."

"Ah, but I can make it fly," boasted Taj.

"You must be crazy," I said, disgusted with him.

"You want to see?" He said.

"Sure," I said. "I would like to see you do that."

"Very well, but not when so many people are around. We will be back around nine o'clock, no one is around, I will show you."

Since it was around four o'clock in the afternoon, I was wondering how we would spend the time, but the Master had his own plan.

"We have plenty of time to do what needs to be done," he said. "I would like you to meet Cheik Al Huseini." This was the first time I met the great man, who later contributed greatly to my studies.

We went back into the car, and drove to the Cheik's house. The house was small and modest, built sturdily of stone, with thick walls. The door was low, as was normal for middle class Arab houses. This style was followed for many years, for the sake of safety and security. Apparently, the conquering Ottomans used to sweep into houses that had large entrances while riding on their horses, and thus be able to kill and destroy anyone and everything inside. The low entrances forced the rider to get off his horse first, making him much less dangerous to the inhabitants.

In the big living room, which they called the *Dar*, many sofas were placed against the walls, arranged next to each other. About twenty to thirty men were present, dressed in Arab robes and turbans. All were elderly, with long white beards. The Cheik was sitting in the place of honor. When the Master arrived, everyone stood up, repeating the word "*oustaz, oustaz,*" to each other, meaning "teacher." Someone pointed at Taj, and said, "The Afrit is already here." I thought this description fitted Taj perfectly, but expected him to be angry. To my surprise, he seemed pleased by being called that name, and grinned at me like a delighted child.

We sat down, and the men came to kiss the hand of the Master. The light was low, only one lamp was turned on, but I could see that one person did not get up from his seat.

Since this was strange behavior, I looked at him carefully, and to my amazement recognized the old Tuareg, whom I had met years ago in the suk in Damascus, the man who was cut in half. He recognized me as well, smiled, and motioned to me to come and sit by him. I came, and he said jokingly, "Don't start searching for the rest of my body..." I laughed, a little guilty, because that was exactly what I was planning to do.

At any rate I could see nothing, since the long robe he wore covered everything. Everyone conversed in Arabic, which by now I spoke very well, and after a while the Cheik motioned most people out. Eight of us remained in the room. The Master, Taj, and myself, were the only outsiders. The Cheik, the Tuareg, and three other elderly Arabs completed the number of the people who were permitted to attend.

At that moment, a man came from an inside room, carrying a big copper pot, full of steaming hot water. He put the pot on a table in front of the Cheik, addressing him by the title *Mawlana*. This title meant "you are a ruler over me," and was used only to address kings, sultans, or prophets. I was surprised.

This title belonged to very important people, but the house and everything in it spoke of middle class. So what could this mean? The Cheik must have been a very important person, somehow. I planned to ask the Master about it later, not wishing to disturb him with questions at the moment, since I was sure strange things were about to begin to occur.

I was sitting near enough to the Cheik to see everything very clearly, and waited breathlessly for the events that were to come. The Cheik took three pieces of blank paper, and threw them into the hot water in the copper container. The room was completely silent, no one moved, except Taj, who whispered to me, "You are going to like what you see, it's fun, but don't move no matter what happens." I nodded, and concentrated on the pot, looking occasionally at Taj for clarification.

Somehow he assumed the role of my guide to the occult world, and I realized he knew exactly what was taking place. "Shush, just look at the container, something is about to happen," he said. I went on staring at the pot.

Suddenly, in a blink of an eye, the water in the container disappeared, and the three pieces of paper burst out of the container. They lined up in the air, without any support, one after the other. They waved about for a few seconds, then merged and became one larger piece. The piece of paper started swirling in the air, rotating around itself, quicker and quicker, and suddenly stopped in mid motion. It was suspended in the air, completely still, and in a flash, letters appeared on it, printed clear, black, and easily visible from where I was sitting, though I could not make out the words.

The Cheik got up, approached the paper, read the words, and then asked one of the people attending to close the shutters on all the windows. The room became very dark, and the words, seemingly separated from the paper, glowed in air like a bright hologram.

The Cheik called Taj, and asked him to read the words. I could not hear what they said to each other, but they seemed to agree on something, as they stood there, nodding their heads. Then Taj came back to me. I asked him, "What was that?" He stepped on my foot to quiet me. His large foot's imprint was painful, so I shut up. Everyone else seemed to accept the phenomenon without trouble, and gazed at the Cheik as he began to move in a strange manner.

He looked to the left, mumbling something incoherent, then to the right, saying the same incomprehensible things, repeating the sequence twice. Then he lifted his hands as if in prayer, in the manner shared by both Jews and Muslims. Touching his chest and pushing his hands in front of him, he said, "*Ahlan, ahalan, ahlan, ahalan, bee salamah.*" The letters were still glowing in front of him in the air, and he added, "*Asma' oo hoosmah ath sab'ha.*"

I turned and pinched Taj, whispering feverishly, "Explain!"

"Don't you know anything?" said Taj. "These are the names of seven Afrit. They are going to open the gate of the underworld for us."

"But..." He stomped on my foot again to shut me up, and it really hurt and I kept quiet.

The Cheik said, rather loudly: "*Bakhooor, bakhooor!*" A man appeared out of nowhere and brought an incense holder. The Cheik moved it back and forth, the room filled with smoke, and everyone started to chant and mumble very loudly. I understood nothing at all of what they said. It seemed they were speaking in tongues, and the effect was frightening.

They went on for a couple of minutes, then stopped abruptly. At that instant, the letters pulled together, became one shining ball of light of intense silver color, and zoomed out of the room into thin air.

One of the people opened the shutters and the late afternoon light streamed in. The Cheik put his right hand on his heart and said "Thank you" three times. I was wondering who exactly he was thanking, and who, originally, was he praying for, since he never used the words God, Allah, or any other recognizable deity name. I did not realize at the time that the Ulema, even when they were Arabs, where not Muslims, and had their own, very different, world view.

The Master got up. Everyone rose with him, their robes swishing and making a faint sound in the quiet room. The Tuareg floated in the air. I looked at him, doing my best to control my discomfort. His upper body was solid, but the bottom half of the robe was obviously empty as it swirled around him, making the absence of his lower body extremely and disturbingly clear. He seemed like an apparition, a ghost.

Everyone came to the Master, bowed to him, and then grabbed his hand with both of theirs, in a way that was clearly ceremonial. The right hand's thumb was hitting the spot between the thumb and first finger of the left hand, and then the left hand covered the right hand. The Tuareg floated near the master and did the same thing. Everyone looked at each other and thanked each other a few times, following their thanks with the words "*Rama Ahaab.*" I did not know this word, and was not aware that they were speaking Ana'kh, the language that was shared by the Anunnaki and the Ulema. And yet I sensed that there was something very special about the way they spoke, as if by instinct. I was staring at the people and trying to understand their words until the Master tapped me on the shoulder and told me to come out.

Taj left with me, and said, "You talk too much. You should be paying more attention, such an occasion is not likely to happen again!" I shrugged, but I had to admit to myself that he was right, these events were probably unique.

To my surprise, I was beginning to like the Sudanese, and no longer felt threatened by his strange appearance and bizarre behavior. As if reading my mind, he put his hand in the inner pocket of his ill fitting and flashy jacket, pulled out two lollipops, and handed me one.

"Won't you tell me a bit about the Afrit?" I asked, licking my lollipop. "I am not sure why we need to call them. Why can't we just go into the underground city? I don't quite understand anything that is going on here."

"In your home, in France, do you have a *Jaras*, a bell, on your door?" he asked.

"Yes, of course," I said, surprised at the question.

"Well, you see, the underground city do not have a Jaras, and it is locked. If you want to come in, someone must let you in. The Afrit can help you, but you have to call them in a special way. Otherwise, they don't know you want them to open the door. How would they know? They are not too clever."

"Where is the door?" I said.

He pointed to the ground. "Under you, under the house, there is a door. Right under the Cheik's house. A door to the *Aboo*, the

deep abyss. It is also called *Dahleeth*, meaning an underground labyrinth."
"Are there other doors?"
"Very likely, but I only know this one."
One of the people came out of the house, motioned to us to come in, and said, "We are ready." In the house, everyone was wearing a white robe, and to my surprise, their heads were covered with the type of head scarf Jews sometimes wore in the synagogue. To confuse the issue even further, one was holding a scroll that resembled a Torah. I felt desperate. Were they going to delay our journey again and start praying? I really wanted move on, see the Afrit, have the adventure. I was tired of the delays. Thankfully, one of them handed me a robe and commanded me to go change my clothes, which I did, but Taj did not change his attire. I asked him why he was not required to do so, and he explained that he was not one of the *Al Moomawariin*, or the enlightened ones, so he was not required to wear the special outfit. This did not really clarify the matter, since I was not one of the enlightened ones either, but I decided to let it pass.
Taj seemed to be right about the door being under the Cheik's house, because we started to descend the steps to the basement. The basement was long and narrow, and had a very high ceiling, perhaps the height of two stories.
Everything, floor, walls, ceiling, were made entirely of gray cement. It smelled of dampness, and was very cold.
We went through a one room after another, all narrow and long, eventually reaching a small room that had an iron gate by its far wall. The Cheik opened the gate with a large key, and behind it was a second door, made of thick wood. A second key opened this one. Suddenly a thought struck me. Why did he need a key? Why couldn't a man who had such supernatural powers simply command the doors to open? Or pass through them like a ghost, for that matter? I expressed my thought to Taj. "It won't work," said Taj. "Yes, of course the Cheik could pass through doors, but how would he take you with him?"
"What do you mean?" I asked, bewildered.
"You are not enlightened as yet. You cannot use supernatural means of transportation at this stage, so if he wants you, or me, for that matter, to pass through these doors, he must take you inside in a normal way. If he tried, you will just bang against the doors and hurt yourself, while he would be on the other side." I began to see that Taj was not stupid at all. Childish, and

sometimes pretending to be silly and play silly games, but deep down, he was extremely knowledgeable.

We stood together in the small room, exactly like all the other rooms in the basement. The Cheik said, "Let the boy be the last one. He needs protection. Taj, come here."

Taj joined him at the front of the line, and we entered a long corridor. As we were walking, the corridor began to shift its shape. I felt seasick, nauseated, my balance was lost. The floor, and walls, everything was moving, rolling, undulating. I did not see clearly, and wondered how long this torment would last, when suddenly all movement stopped.

I looked around and nearly jumped with terror.

The simple corridor became a cave! A natural cave, not a man made structure. Stone, dirt, and natural formations were all around me. It smelled damp and filthy, water were oozing from some of the walls, and the light was dim. I did not like the place.

The Master told everyone except me to stand in a crescent shaped row, and hold hands. He ordered me to stand behind the crescent, and not to touch anyone. I was hurt. I felt neglected, as if I were not part of the group, until one of the people turned to me and said kindly, "Don't be upset, my boy. This is for your protection." So I just stood there behind the people, feeling silly in my long white robe, but not unhappy anymore.

At that moment, Taj made a sweeping motion with his hands and body, and screamed a few words. The horrible sound he emitted was not human. It was very likely the loudest sound I had ever heard. He continued to move his hands violently, grabbed some dirt from the ground, and threw it up in the air. He pronounced a word that to me sounded like a name, and followed it by the word "*Eehdar*!" three times. Then he said, "*Oodkhool*," three times. Immediately, a rubbery kind of form moved to the left, changed to a paste-like substance, and attached itself like glue to the wall. The sticky, pale mess looked
like ectoplasm. Taj repeated his actions a few times, manifesting a new ectoplasmic manifestation on the wall with each call. Then, he looked at the Cheik and said, "*Tamam*!"

The Cheik and Taj were engaged in a conversation in low voices. They seemed to be in agreement, since the Cheik said, "Yes, go ahead." Taj advanced toward the ectoplasmic forms, put his hand in his jacket's pocket and took something out, and gave some to each of them. At this moment, the Cheik stepped forward, ready to take over, and said "*Iibriiz*!" The forms burst into flame, which burned the ectoplasm and produced a thick fog. From the fog appeared human forms, but there were only six of them. The Cheik said "*Wawsabeh!*" The Master came forward, stood by the Cheik, and the Cheik repeated the word, adding, "*Anna a'mooree khum!*" and the seventh creature came.

Later Taj told me that these Afrit were originally created by the Cheik for a reason, as they usually are, and in the normal state of events were supposed to become the Cheik's loyal servants. However, the Cheik made a mistake and did not perform the exact requirements needed in the procedure of the creation, and therefore he lost control over the Afrit. The result was disturbing. The seven Afrit developed independent and rather evil habits, and did not quite obey the Cheik as they should.

For some reason, the only one who could call them to appear was Taj. However, that is all he could do. Since Taj was not an Enlightened One, he could not control them once they came, and to a certain extent was at their mercy and had to have an Ulema present if he were to avoid potential harm.

As for another Ulema controlling them instead of Taj and the Cheik, that was not possible. The Ulema have four categories, based on their form of existence. Some Ulema are physical and live as humans, like the Master and the Cheik. Some used to be physical, but were no longer so. Some, like the Tuareg, straddled both forms. Others have never occupied a human form. All four versions of the Ulema can exercise immense powers, no matter if they are physical or non physical, but a physical Ulema can only control non physical entities, such as these Afrit, if he was their creator.

I shuddered as I watched the Afrit. At this point of my studies, I had my share of supernatural incidences, but I have never been so shaken before. In the semi darkness of this miserable, damp place, the Afrit were truly terrifying. Each had a more or less human face, but in this almost normal face the eyes were not at all normal. Instead, each Afrit had two circular orbs, with white background and a black pupil that stood out as if painted. The eyes did not move. If the Afrit wanted to look to the side, it had to move its whole head. The head was not connected to the body. Instead, it floated in a disconcerting, eerie fashion, just above the body. When the Afrit manifest, their bodies often appear first and for a few minutes appears headless, until they choose to manifest the head.

This fact, coupled by their appalling ugliness, can frighten a human being to the point of death. There had been recorded incidents of people dying of heart attack or stroke caused by such events. I kept myself as calm as possible and continued to study the Afrit. The heads were bad enough, but the bodies were even worse.

They were tinted a shadowy, ugly, dark color. The torso resembled the shape of a bat. Their arms were attached to the back of the body, and the hands had extremely long fingers. Since the Afrit don't eat or breathe, they don't need a stomach and a diaphragm. Therefore, the body had a sort of visible cavity in the front, where these organs would have been. The legs were twisted, like entangled wires, which must help the Afrit as they jump. They rarely stay in one place for long, and keep shaking and moving and twitching. They looked back at us, their ugly faces twisted in a devilish, vicious smile. They kept chattering among themselves and pointing at us with their long fingers. But Taj told me that despite their apparent boldness, they were afraid of the Enlightened Ones. Any Afrit can see the shining auras of the Ulema, and for some reason they are terrified of these auras.

The Cheik commanded the Afrit to open the door. I did not understand the language he spoke, but I figured it out because he used the word "*Babu,*" which is so similar to the word *Ba'ab*. Babu is really a door, though, while ba'ab is a gate, but the words were close enough to make it clear to me that they were going to open the door to the underground world. I was speechless with anticipation. Everyone stood still, looking at the far wall of the cave, so I stared at it too, not knowing what to expect.

The far wall of the cave suddenly collapsed, in total silence. It felt like a silent movie, because there was no dust and no sound of falling stones during the procedure.

The stones tumbled down quietly, one by one, disappearing altogether rather than forming a solid pile.

The wall was replaced by dark, hazy fog that allowed us a glimpse of some far away buildings. "Now," said the Cheik to Taj, "Let's follow the Afrit, but don't let them play tricks on you." Taj nodded. We went through the fog, following another corridor and crossing identical rooms that seemed to follow each other in succession, all the while seeing the far off buildings in the distance.

The Cheik started reciting something. The Afrit were jumping up and down like carousel horses, while pushing forward with great speed, and were already a good distance away from us, going on their own mysterious errands. Taj said to me, "You can now move to the front, it's safe now, the Afrit won't pay much attention to us anymore." I quickly moved near the Master at the head of the line, and no one took notice of what I was doing. We did not move on yet. The Cheik asked Taj to show him a piece of paper he was holding, probably a kind of a map, and asked, "Do you know which room we need?"

"Yes," said Taj. "I know exactly where it is, it's very near us. I will go in, and if I find something, I will bring some pieces back to you so you can see them, and then we can all go in and bring everything."

Taj left for about five minutes, and returned with a beautiful pearl necklace, a few diamonds, and some Phoenician coins. He told the Cheik and the Master, "We can go in now, but remember, you promised that all the gold belongs to Taj."

"Of course," said the Cheik casually. "But remember," said the Master, "We are not just going into the treasure room. You will also take us to the other room, as you promised."

It was clear to me that the Ulema were not in the least interested in the treasure, but there was something else in this underground cavern that meant much more to them than any gold or diamonds. The Ulema do not need gold.

They can manufacture whatever wealth they need, and they never manufacture or acquire more than they need. Riches are of no interest to them at all.

"Certainly I will take you to the other room," said Taj. "I know exactly where it is." He seemed quite pleased by the bargain.

We followed Taj into a small, closed room. It had no windows but was brightly lit, allowing us to see gold, gems, diamonds, and pearls stashed in boxes, jars, or simply thrown on the floor in heaps. However, I was not very interested in gold either. What I wondered about was the source of the mysterious illumination. No windows, no lamps, no candles, but bright light in every corner of the room. What could cause this? Suddenly I realized it had to be the same type of light that was discovered in the Pharaonic tombs and catacombs of ancient Egypt. Originally, the archaeologists who went there were baffled by the light in the Egyptian tunnels, until they discovered the contraption that the ancient Egyptians had created. They found conical objects that functioned like modern batteries, producing light that was so much like normal electrical light that there was hardly a difference. The batteries had to be placed in a certain way against each other, or they would not light, and worse, could burn the user since they packed a lot of energy in their structure. I suspected this had to be the same type of illumination.

Taj pointed the door that would take us to the room the Master wished to visit. The Master asked him, "Do you want to come with us?"

"I will follow you as soon as I am finished here," said Taj, grinning. He pulled some linen bags from under his jacket, and busily started filling them with the treasure. The Master smiled indulgently at him, as if Taj was a child playing with some toys that meant little to adults but pleased the child a great deal. He said to the rest of us, "Well then, let's go to the next room." We opened the door. Inside it was pitch black, but the Master stepped in without the slightest hesitation, and we followed. I envied his confidence. As far as I was concerned, how did we know an angry Afrit was not waiting for us? But since no one else showed any fear, I went with them. We could see nothing, but the Master kept talking to us and so we were able to follow him. All of sudden, bright light filled the room. I blinked a few times, and then saw the Master standing by one of the walls, holding two conical, golden objects in each hand, positioned against each other. I was right, here were the ancient batteries.

The room was empty of furniture other than a beautiful wooden table, carved into arabesques, much like Moroccan furniture. The Master placed the batteries carefully on the table, making sure the alignment allowed them to continue to produce light. I looked around. Other than the batteries and the table, the only object in the room was a large Phoenician urn, standing in one of the corners.

"We are going to leave you here for a short while," said the Master to the group. "The Cheik and I are going to get the materials we need for our project."

"We'll be right back," added the Cheik with what seemed to me rather misguided optimism. There were Djinn and Afrit here! Wasn't anyone concerned about these devils? The Master and the Cheik walked to the end of the room, very slowly, with measured, matching steps, as if choreographed. Then they reached the far wall, and literally went through the wall to the other side. I was not exactly shocked, since I have seen the Master go through walls before.

It is an interesting phenomenon, but not as mysterious as one might think. To put it simply, the Ulema know how to control molecules; the Master had explained it to me thoroughly. Everyone knows that there is plenty of empty space between the molecules of any matter, and the Ulema make use of that fact with a specialized procedure.

As the person who wishes to cross approaches the wall, the wall gradually becomes soft, as if its molecules fragment themselves, and the human body simultaneously does the same.

The spaces between the molecules of both grow and readjust. The person and the wall keep their shapes for an instant, then their molecules mingle and allow the passage. At that moment, the person passes to the other side, the molecules separate, and both wall and person become solid and normal again.

The rest of us waited for about half an hour. I was beginning to worry. The Cheik said they would be right back! Something must have prevented them from doing so. Perhaps the Afrit, who have by now completely disappeared, took them away, kidnapped them, led them

somewhere horrible? I asked some of the other people if they knew what was going on, but they had no idea where the Cheik and the Master went. However, they did not seem worried, making it clear to me that they trusted these two to know what to do. "Don't worry," one of them said to me. "They can handle a lot worse than those stupid Afrit."

"I don't wish to contradict, Sir," I said, "but these Afrit seem pretty dangerous to me. The way they were pointing and smiling..." The others laughed. "I have seen the Cheik and the Master handle much worse entities," said the man who spoke to me, very kindly. "Remember, the Afrit are cowards. They are mortally afraid of the auras of the Ulema."

"But I understand the Cheik needs some help because of the way he handled their creation," I said.

"Yes, this is true," said the man. "These Afrit did turn out a bit wild. But with the Master there, they will never dare to harm them." I had to be content with that. So I went in search of Taj, to see how he was doing with the treasure, perhaps help him finish filling his bags. I called him and was about to reenter the room, but I heard him scream, "Don't come here!" and he tumbled out of the room, bleeding, and slammed the door behind him. "The Afrit beat me," he gasped. "Beat me very badly."

"But Taj, you could handle those seven Afrit so well! What happened to give them power over you?"

"Seven? Are you joking? There is a colony here, something like forty of fifty Afrit, and they all rushed at me and would not let me take the gold."

"Is it their gold?" I asked. "What do they want it for, anyway? They don't need money."

"No, it's not their gold. It used to belong to the Phoenicians, and now it belongs to no one in particular. But the Afrit like to play with it. They like shining things."

"But you are holding one bag, I see."

"Yes, I managed to save one bag. They got all the others, those slimy devils." He smiled, regaining his composure. "Never mind, though. After all, I will be a very wealthy man even with just one bag. This treasure is amazing...Anyway, we must secure the door. Hold the bag for a minute." He pushed the bag in my hands, turned, and repeated the same words he used when he originally called the Afrit, and gestured in the same way. While he was doing that, I heard shrieks and screams, which he later explained was the way the Afrit spoke as they were chased away. "That is that," he said, surveying the door with satisfaction. "They won't bother us again." He took the bag and smiled at me through the caked blood and filth on his face. "A successful treasure hunt, ah, Germain? And some day I'll come back for more."

Back in the other room, I saw, to my considerable relief, that the Cheik and the Master have returned. The Cheik was holding a stack of forty or fifty sheets made of shiny plastic, or plasma, or glass, and the Master had the same size stack, but of a different type of material, brownish yellow like corn.

"What is that?" I asked Taj.

"I have no idea," said Taj. "They only told me which room I was supposed to take them to, but they did not tell me what project they were engaged in. I must say
I have a hunch it is something terribly important." I thought so too, since the Cheik and the Master seemed to be extremely solemn, and everyone else was completely silent. There was a strong feeling of expectation in the room. They each put his stack on the table, the Cheik on the right, the Master on the left, leaving a space between the stacks, and I noticed that the space matched the size of the stacks. The Master brought the urn from the corner to the table, and made a motion of pouring something out of the urn into the space between the stacks. I saw nothing

coming out of the jar, but I figured that it might be an invisible substance.

This went on for about twenty seconds, then the Master returned the urn to the corner. The Cheik took one sheet from his stack, and put it in the space between the stacks. The Master then took a sheet from his own stack, put it on the Cheik's sheet, and waited a couple of seconds. Then the Master flipped his sheet back side up, and to my absolute amazement, there was print on the sheet, strong and black, consisting of strange symbols and letters I did not recognize.

Piling the sheets on top of each other, they did the same to all of them. Surprisingly, the stack, when finished, was reduced in size to about a half of the original sheets, even though I could not see it reducing itself while it was worked on. I think that the plasma sheets were absorbed into the corn-like paper as the print was produced, but I am not sure. The Cheik pulled out a silk scarf from his robe, put the stack on the scarf, rolled it, lifted the ends of the scarf and tied them together, all in a ritualistic way. Then he said, "*Al Hamdu*" twice.

They turned to go, and we left the room. The Master, throughout the entire time, paid hardly any attention to me, which bothered me a little.

I felt neglected, even abandoned. He must have noticed my unhappy face, because he put his hand on my shoulder, took me back into the room, and said, "Look!" To my amazement, the room was entirely empty. The table and the urn had disappeared. I was confused and uncomfortable. I could not understand why all that was necessary. Why Afrit? Why those doors? Where did the table go? What was this document and why was it worth all this effort? He laughed at my questions and said, "Look at the wall."

The light was dimming as we spoke, and finally disappeared. It seemed this adventure was over, and I said, rebelliously, that I wish things were made clear to me, because otherwise, I have learned nothing.

"I will explain everything later, Germain. I promise"

"But what about the city you said we are about to see? The city where the Founding Fathers of the Ulema used to come to? The city from before the Deluge?"

"So you want to see more? This was not enough?"

"Yes," I said. "Basically, all I saw was you and the Cheik going through a wall and Taj fighting with the Afrit, which I admit were scary but were not too significant, I believe. I did not see anything remotely connected to the ancient city."

"Well," he said, "in this case, turn, and walk with me. You are already walking in this city."

I looked around, and saw nothing, but he said, "Keep walking, it will come."

I should have trusted him more fully. After all, when did he ever disappoint me? I felt remorseful as the miracle began to enfold in front of my eyes, but thankfully, he did not hold my short term rebellion against me, and went on cheerfully enough. Slowly, the ancient city started to appear like a Polaroid picture in front of me. The colors of the city were such as I have never seen before, glowing colors of incredible beauty. The Master explained that this was because the city was located in a space that had the same temperature everywhere, and no pressure on any object. Unlike earth.

"What do you mean, Master, when you say 'unlike earth' like that? Are we not on earth?"

"No, we have left earth when the Afrit opened the door and made the cave wall collapse. We are now in another dimension," said the Master. "Everything looks a little different here."

The city became clearer, and I thought it looked like a holographic projection, either from the past, or from the future. The buildings, though beautiful, had a sense of alien, remote places. We were now walking in a well-illuminated street, the windows of the buildings shining with lights as well. The air was soft and fragrant.

"I see buildings and streets," I said. "But where are the people?"

"They are here, but they are invisible to you. Your eyes are not constructed to see them, not yet," he said. "Well, it is time to leave. Let's go up these stair." We started climbing a very high, stone stairway that led from the street into a destination that was not quite visible.

I was surprised that we were not retracing out steps into the Cheik's house, but the Master said there was no need for that, and that exits were available in various locations, and not as difficult to achieve as entrances. So we climbed the stairs, and when we reached the top, I saw a huge gray wall on my left, and noticed that the pavement turned into sand. The huge gray wall was the side of the Anunnaki stone. I understood that we exited from a hole under the big stone, were out of the strange dimension, and back on earth.

"So that is what Taj meant when he said he would make the stone fly?" I said.

"Yes, a rather poetic way of describing our trip," said the Master.

"Master, I am not wearing the white robe! I am wearing the normal clothes I left at the Cheik's house."

"Indeed, and so is everyone else," he said, pointing to the rest of the company, who were already standing near the giant stone, and wearing normal clothes.

"So what did we come here for? Surely not just to give Taj his treasure?"

"We came for the book, Germain. Everything we did was much worth it, even the encounter with the unpleasant and stupid Afrit. We have recently heard that the book was here, in this dimension, after having searched for it unsuccessfully for generations. And now we have recoverd a copy of the most important book in the world."

"The strange book you printed from the stacks? What is it?"

"It is one of the very few copies in existence of what is probably the oldest book to have ever been written. A book the Anunnaki had valued very much. It is called *The Book of Rama Dosh*."

I didn't know why, but a shiver went through my spine when I heard the name of the Ancient book; the sound of the name triggered a reaction in my mind. For a second I had a feeling of tottering on the brink of a dark, warm abyss that contained something older than the universe, and glowed with endless stars. It passed quickly, and the Master continued.

"In the future, you will have the privilege of studying it. It contains the knowledge that may, some day, save humanity from its own folly. At least I hope so with all my heart. And now, back to Damascus! Our friendly driver is waiting for us in the car."

Hawaa: The air. From Hawaa derived the Phoenician word Haw, which means air. From the Phoenician, derived the Aramaic word Hawa, which means air, and from the Aramaic, derived the Arabic word Hawa' meaning the same thing.

Hawwah: *Aramaic/Hebrew/Anak'h/Arabic/Akkadian/Noun.*
Original name of Eve. The Anunnaki woman who created the human race, more precisely the 7 prototypes of man, according to the Anunnaki-Ulema's Book of Ramadosh, and other esoteric manuscripts. The Bible called her Eve, "however what the Bible told us about her is totally wrong..." said Ulema Al Huseini. The

Akashic records of the Anunnaki reveal that Hawwa was the Angel Gb'r, known also as Gib-ra-il, (Angel Gabriel), the guardian and governor of Janat Adan (Garden of Eden). Eve is also closely associated with the Sumerian high goddess Ninhursag through her epithet "Mother of All the Living."

In old Hebrew and Aramaic, Eve is called "Chavvah", and "Hawwah".

In Arabic (Pre-Islamic, Al Jaheeliya Years, and Islamic era), and Syriac, Eve is called "Hawwah". Dr. Rohl agrees that Hawwah (Chavvah), was mistakenly called Eve by generations.

The Hebrew name Hawwah links this very rare name with the prime verbal root *"to make live"* - hayah - which is in itself an Akkadian word. According to the Anunnaki, Eve with the help of the Anunnaki created Adam.

Contrary to all beliefs, including what Judaism, Christianity and Islam teach us, Eve was not created from the rib of Adam. Men were created from an early female form that was "fertilized" by the leaders and the elite of the Anunnaki.

They lived in quarantined cities, and had both sons and daughters fathered by the Anunnaki. According to the Book of Ramadosh, at the beginning, Hawwah, the woman was created first, not man. The early women were called "Women of the Light"; they were the early female-forms on earth.

Early humans who lived during that era called the quarantined city of these women "The City of Mirage", and "The City of Beautiful Illusion," since the most attractive women from earth lived there.

And the quasi-humans who were made out of earth were not allowed to interact with these women. Thousands of years later, the inhabitants of what is today the Arab Peninsula and the lands bordering Persia, the United Arab Emirates, and India, called these women "The Women of Light", and those who were allowed to "mix with them" were called "The Sons of Light". From this early human race, all humans came to life.

Contrary to all beliefs, including what Judaism, Christianity and Islam taught us, Eve was not created from the rib of Adam. Men were created from an early female form that was 'fertilized' by the leaders and the elite of the Anunnaki. The women lived in quarantined cities, and had both sons and daughters fathered by the Anunnaki. Ramadosh Epics of Hawwa. The Anunnaki-Ulema maintain that the universe was created from a molecule smaller than the tip of a pin, taking less than three seconds. The language is metaphoric, the science is highly visible – much like our own Genesis whose language covers the Big Bang and the Theory of Evolution. And Hawwa existed at the dawn of the Genesis.

Transliteration of a text from the Book of Ramadosh

1. Inna bida rama dosh kali kilma
wa falki uzzu ina wa anru dani (Dounia)
2. u rama dosh khalki shama u erdi
3. wa erdi naya shak-lu fari mara anu absi
u rama dosh liwa basra erdi
4. u rama dosh shadah ilmu erdi rou'a min bashri
5. u rama dosh khalka belti isama shavah
6. wa leilu wa fagru subhi yomou badri.
7. u hawwa marki-ya kila la-ma nazri. U rama dosh kali na inna erdi wadoo kourba shamsi, wa noura khalku, wa noura barku. u hawwa basri noura gulba.
8. u hawwa ma dari akhlu jisma ma khalki sartu inaya mayi, rama dosh kali da jamu ma'aa rama faku erdi wa zahra erdi u hawwa basri noura gulba.
9. u hawwa ma dari ma'uu u rama dosh daa'ghasbu ma'ii inna boukari hawwa nasmu-ya, w hawa'u nafsuru, u hawwa basri noura gulba.
10. wa leilu fajri barku itani yomu.
11. u hawwa isha maraadu rama dosh kali na inna erdi khalka ishbu wa fakha zahri gensu u hawwa basri noura gulba.
12. u hawwa na gismu kilu ala tadri abani erdi wa harka nazri kulu ma'aa wa h'azru alama erdi. u hawwa basri noura gulba.
13. u hawwa isha maraadu itani u rama dosh zahru jasru i-ya rim aspsi-nama. Maraadu aliha itani faku erdi hayah lawida, u rama dosh ilmu i-ya haki. U rama dosh kali nama gubla inna hima nama eisha lawida na khalku bashru iina haya-ti
14. wa leilu fajri barku silsu yomu.
15. miba hawwa aspi-nama rama dosh akhza mina jisma-ya wa tourba min erdi abba ma'aa jam'uu inna taboura wa jalsi hawwa taboura nasbu nefsu illa zahru bashru ma innu jismu misla hawwa wa rama dosh ilmu na gulba.
16. u rama dosh isbhahu zakar nami wa uli marku inna ajla bashru na zahru hawwa jisma baadi. U rama i-shem hu Zakar u rama dosh antaka li jalsu wu Zakar jalasi doughra.

17. u rama dosh antaka hawwa la jalsa wu Hawwa basra basharu wa ulma noura gulba.
(Source: The Book of Ramadosh.)

Translation of the text from the Book of Ramadosh

1. In the beginning, Rama Dosh spoke the Word and the universe burst into being and was ready for life.
2. And Rama Dosh created the heaven and the earth.
3. And the earth was without form, and void, and darkness was upon the face of the deep. And only Rama Dosh could see the earth.
4. And Rama Dosh wanted to know what the earth would look like if it were seen by humans.
5. And Rama Dosh created a female human from their own essence, and called her Chavah. In their own image, in the image of Rama Dosh, created they Chavah.
6. And the evening and the morning were the first day.
7. And Chavah was confused, and said, I cannot see. So Rama Dosh said, I shall position the earth not far from the sun, and there will be light: and there was light. And Chavah saw that it was good.
8. And Chavah was not hungry, since her body was not yet complete, but she was thirsty. So Rama Dosh said, Let the water under the heaven be gathered together unto one place, and let the dry land appear: and so it was, and Chavah saw that it was good.
9. And Chavah could not drink, so Rama Dosh made the water go up in steam so Chavah could breathe it, and that was the air, and Chavah saw that it was good.
10. And the evening and the morning were the second day.
11. And Chavah was bored. So Rama Dosh said, Let the earth bring forth grass, the herb yielding seed, and the fruit tree yielding fruit after his kind: and so it was, and Chavah saw that it was good.
12. And Chavah, for her body was not as yet complete, could fly all over the earth. And she moved upon the face of the water and the earth and all the green things. And Chavah saw that it was good.
13. And Chavah was bored again, and Rama Dosh were angry with her and made her sleep. And while she slept, still they realized that she was bored because she was all

alone upon the earth, and Rama Dosh knew that she was right. And Rama Dosh said, it is not good that the woman should be alone. We will make her a help meet for her.

14. And the evening and the morning were the third day.
15. And while Chavah slept, Rama Dosh took a part of her body, and parts from the dirt of the earth, and parts of the water, and mixed them into clay. And they put the clay next to Chavah, and they breathed upon the clay, and it became a man, but he looked like Chavah, and Rama Dosh knew that this was not good.
16. And Rama Dosh pointed Their finger at the sleeping man, and They touched him, and the man changed and no longer looked like Chavah, but like a man. And Rama Dosh named the man Zakar, and commanded him to wake up: and he woke up.
17. And Rama Dosh commanded Chavah to wake up, and she saw the man, and she knew that it was good.

(Source: Translation by Maximillien de Lafayette. From his book "The Anunnaki Ulema Forbidden Knowledge.)

Haya-Saadiraat: Products; results, production. The word Saadiraat is used in Persian/Iranian and Arabic bearing identical meaning. However, the general meaning is a country's national production, and/or what a country, its industry, market, finance, banking industry and commerce produce. One of the largest Iranian banks in the Persian Gulf and the United Arab Emirates region is called "Bank Saadiraat Iran". In Ana'kh, the meaning emphasizes the global production of materials and tools related to health and longevity. Many of these tools are used in DNA laboratories to extend life, alter DNA, and duplicate copies of Anunnaki individuals. Haya-Saadiraat is composed of two words:
a-Haya or Hayah, which means life;
b-Saadiraa, which means tools; products; means. In Arabic, the word Hayah or Hayat means life. It meant the same thing in Proto Aramaic and Hebrew.

Hazi-minzar "Mnaizar": A small but sophisticated device used by the Ulema of the 8th degree to read and decipher codes and symbols from the Book of Ramadosh. It is composed of two words:

a-Hazi, which means to read; to decipher a code.
b-Minzar, which means an observation tool. The word Mnaizar" is a diminutive of Minzar, referring to a smaller Minzar.
Ulema Mordechai gaves us a rare description of Hazi-minzar. His description appeared for the first time in the West in the book "On the Road to Ultimate Knowledge", co-authored by Maximillien de Lafayette and Dr. Ilil Arbel. Here is an excerpt from this book: The scene: Ulema Cheik Al Huseini talking to Germain Lumiere, a young Ulema, who is visiting the Cheik by invitation of Dr. Farid Tayara, a noted Ulema and head of a Masonic Lodge. The place: Baalbeck, Lebanon, in the house of the Cheik.The year: Around 1957-1958.
The text: Germain Lumiere telling the story in his own words:
"Slowly, thoughtfully, I went to Dr. Farid's office, musing on all that has happened to me in the last few days; it was hard to digest.
I was going to get instructions as to when and where I would be given the honor of reading *The Book of Rama Dosh*. Another miracle will manifest in my life. Dr. Farid informed me that the arrangements have been made, and that the next day he would pick me up very early in the morning. We were to drive to Baalbeck, to see Cheik Al Huseini, my host during my previous trip to Baalbeck. It was there that I saw the startling printing of *The Book of Rama Dosh*, in the underground city. It would be nice to see him again.
Dr. Farid added that Ulema Ghandahar, an expert on *The Book of Rama Dosh*, would join us at the Cheik's house.The Cheik, as hospitable and pleasant as ever, was delighted to see me, and hugged me enthusiastically in the friendly and warm Arab fashion. "I knew you had the making of a great Ulema in you, Germain!" he said, holding me at arm's length and looking at my face with great affection. "You were such an attentive youth, and so fearless during our meeting with the Afrit, we were impressed!"
"I wish I had known you were impressed at the time," I said, laughing. "I felt like such a fool, and Taj made fun of me."
"Ah, that is just Taj," he said indulgently. "Such a silly man, like a big baby... But we all love him anyway. And he is doing very well now, with all the gold he got at the underground city."
"He was badly beaten for it by the Afrit," I said.
"You pay the price for everything in this world," said the Cheik philosophically. "But in the end, everything is as it should be. As

we Arabs say, *Machtoob*! It is written... But come in, come in! Ulema Ghandahar is waiting for us in the library." My excitement at the thought of finally reading *The Book of Rama Dosh* hardly needs to be described.

We entered the house and went directly to the library. It was a much smaller room than I expected, and the pretty, carved and glassed over bookcases seemed to contain scholarly, but ordinary books, the kind you would find in any scholar's library. I was surprised, since I expected a huge collection at Cheik Al Huseini's library.

Little did I know what was to come... The Cheik introduced me to Ulema Ghandahar, who shook my hand and said that he would be so happy to acquaint me with the most important book in the world. Cheik Al Huseini went to one of the bookcases and pushed a hidden button among the carvings on the wood. The case swerved to the side, and a short secret passage was revealed. We walked through it to a wooden door, and entered a library of immense proportions.

The ceiling was very high, about fourteen feet in my estimate, and the room stretched to the proportions of a hall. Bookcases lined the walls, floor to ceiling, and more books were stacked on tables. These books were mostly very old, as you could tell from the leather and cloth covers. However, not only books were there. Through the glassed doors on some of the cabinets I saw a huge collection of ancient rolled-up scrolls. There was a divan on one side, and a few comfortable chairs, all done in the sumptuous Arab style.

Diffused light came from the partially covered windows. This was exactly like the library I had imagined Cheik Al Huseini would have. Of course, I thought. There are things here that should never be seen by the non-initiates. He must keep it secret. Cheik Al Huseini went to one of the bookcases, looking for something, and without turning his head said, "Please, help yourselves!" I looked at the table before me, on which three cups of tea, which were not there a minute ago, suddenly materialized, accompanied by some pastries.

I smiled and looked at Dr. Farid, pointing silently at the tea cups. "This is only the beginning," he said mysteriously. My excitement mounted, I could not wait to see *The Book of Rama Dosh*, and I was wondering if that was what the Cheik was looking for. I sipped my tea and took a pastry. It was interesting, I thought, how different the Ulema of the Middle East were from the

Western ones, or the Chinese, even though their goals, aspirations, and ethics were exactly the same.

For example, Rabbi Mordechai always said, "If you can do something normally, there is no reason to use the so-called supernatural powers." Master Li was exactly the same. I was taught the techniques that emphasized the power of mind, not techniques that had the touch of the magical. The Middle Eastern Ulema did not think in those terms.

They comfortably used all the magical techniques they wanted, and in addition, seemed to have contact with non-human entities who lived with them and worked for them. I decided that the people of the Middle East loved emulating the sumptuous style of King Solomon, with his Afrit, gold, talking animals, flying carpets, and rivers of wine.

The Western Ulema tended to work like scientists, with a tendency toward austerity and a simple lifestyle. The differences were dictated by personality and culture, I suppose, because all of them wanted and achieved the same objectives, only reaching them by different roads.The Cheik turned away from the book case, and walked a few feet toward us. He did not find the book he looked for, I thought, worried that it was lost and I will not be seeing it after all. A sense of disappointment went through me, but I noticed that the Cheik was doing something strange. He turned toward the bookcase, lifted his arm, and pointed at the book case.

Then he stopped, not moving. A second later, a book came floating toward him, and hovered in midair. The Cheik sat down and spoke a few words in a language that I did not know, but from the way he said it, I deduced that it was a code. The book floated further toward him, and settled gently on the table. It was a big, heavy book, with a wood bark cover that had no marking on it to show what was its title. The Cheik did not touch it. Instead, he went to a small table on the side of the divan, and brought a small box made of dark wood, inlaid with silver and mother of pearl. He put it next to the book.

"Germain, would you please go to the bathroom next door, take a shower, and put on the white robe that hangs on the door," the Cheik said. "We'll wait for you." I did as I was told. While showering, I wondered if the book on the table was indeed *The Book of Rama Dosh*. How could it be? It looked quite different when the Cheik and Master Li printed it with the help of the Miraya plates and the light. Then, it looked like shining plastic, very modern, while the book on the library table was a normal, old book. Later I found out how this worked, so I might as well explain it right here. *The Book of Rama Dosh* exists as only one copy. It is located in another dimension. Each time an Anunnaki-Ulema needs a copy, it must be printed directly from this original.

Calling it requires special situations and techniques, such as I have seen in the underground city, but the advantage is, each copy is an exact facsimile of the original. Other ancient books are subject to mistakes in printing, incorrect interpretation of words, etc., but not *The Book of Rama Dosh*. If this was the same copy that was printed in my presence, then the Cheik took the plates, which I remember him to wrap carefully in a silk scarf, to his own library, and there made sure it is properly wrapped in wood bark. It would never be wrapped in leather or any other animal-related substance.

Of course, I could not be sure that this was the same copy, but no matter what, the content was always identical to the true, the one *Book of Rama Dosh*. I put on the white robe, returned to the library and sat at the table with the other three. Cheik Al Huseini opened the book, so now I knew that must be *The Book of Rama Dosh*. I tried to keep calm. This would be the first time I would see Ana'kh printed in a book! And who could tell what the book is about? The Cheik turned the page. It looked old. He turned a few other pages, each looking newer and smoother than the last. None of the pages had anything written on it, though. And yet, the three others seemed to be absorbed in reading the book! Was I going mad? I did not want to interrupt them, or ask questions, but I was beginning to feel desperate. Another page was turned, and it was again completely blank. I sighed with irritation. The Cheik suddenly stopped, looked at me and said, "*Moo Akhazaa*, forgive me, please." He laughed gently.

"You cannot see the writing without the necessary machine," he continued. "We no longer need it, at this stage, and when you get to stage 18 and over, you won't need it either, but for the moment, this machine will help you see the writing." He opened the little box that was on the table next to the book, and took out a sophisticated-looking contraption. It was obviously meant to be used as eyeglasses, but did not look like modern ones. Rather, it was more like a Seventeenth Century Swiss watch, and I saw wheels attached to it on which certain letters and numbers were written, some big, some small, in an old and elegant font, looking like codes. "What is this?" I asked. "It is going to help your vision," said Cheik Al Huseini.

"Take a look at how it is constructed." There were three layers of lenses for each eye, made of glass or crystal, completely transparent. A small wheel, made of gold and edged with green topaz, was attached to each lens, all on one side. Each wheel had a little knob used for adjusting the codes. You would lift each lens individually, and adjust the wheel to the required code. On the other side was a larger wheel, about twice the size of the little wheels, and it adjusted itself to the position of the small wheels once they were in the perfect position.

Once the arrangement of the lenses and wheels was complete, the machine would allow you to see colors we usually do not see on earth. Within these colors reside separate dimensions, or perhaps the colors reside in these dimensions, which is really one and the same. It is as if a door is opened to a spatial gate, an entrance to these parallel dimensions. You are on earth, but through your Conduit, you are entering an unearthly, separate dimension.

"Now, put the machine on, and look at the bookcases. Don't look at the light from the window. This will allow your retina to adjust, and will bring up certain visual faculties."

"What does it do?" I asked, putting the machine on.

"It emulates the natural vision of the Anunnaki, who do not possess a retina, but a more complex mechanism. Even if you close your eyes, once you put the machine on, you can still see."

"What is the name of this machine?" I asked, still looking at the bookcase, as directed.

"It is called *Minaizar*, which is a diminutive of Minzar, the ability to see. The vision through the Minaizar is called *Nazra*." Said the Cheik.

"I am seeing something strange," I said. "The bookcase is suddenly huge, astronomical..."

"But it is still visually very clear, right?" said the Cheik. "Unlike the usual type of visual enlargement, like a magnifier, which blurs everything and forces you to step back, the Minaizar retains its sharp image."

"This is true," I said, "But I feel a little dizzy..." I closed my eyes to refresh them, and was amazed that I could still see, just as the Cheik said before. I opened my eyes and returned to the table, sat down and looked at the book through the machine. Geometrical forms and numerical symbols were printed on the page I looked at. As I was gazing at them, they opened up, unfolded, and I saw letters coming through and appear on the page. Everything was written in pure, original Ana'kh. I could read *The Book of Rama Dosh*!Here I must explain a few things about Ana'kh, which would clarify my reading.

Ana'kh is a unique language, and has some characteristics that no earthly language possesses. For example, when one wants to translate a page verbally from say, Latin to English, each person will have slight variations on the text that they will produce. The same would happen in simultaneous translation of any living language by a translator in the United Nations.

Even when translating a book on paper the variations will appear, which is making translation more an art than a science. Not so with Ana'kh. If a hundred Ulema will verbally translate a page written in Ana'kh, they will use the exact same words, in any language they use. The same goes for written translations. They are not really translating.

They are transmitting, rather, with the help of the Conduit, and no variations will ever occur. Another interesting trait is that the phonetics make themselves clearly "heard" as you read Ana'kh, even if you have never seen or heard the word you are reading. The words pronounce themselves for you, and no mistakes are ever made.

The machine, of course, facilitates that, but it is accomplished by the Conduit. The machine is actually linked directly to the Conduit.

In any book, you cannot start in the middle of a paragraph or a word and still know what the page is all about. You must read a certain amount to grasp the meaning. With Ana'kh, each word presents its own meaning and message. There is no need for grammatical sequences. The words, helped by the machine, follow you, rather than you follow them. In an ordinary book, you have to go back to certain pages if you want to retrace something. In Ana'kh, because of this tendency of the words to follow, you don't need to go back. Rather, you call the word to you.

The simplest analogy would be a search engine on a computer. You type the word on a search engine, and the connected messages appear. That is what happens with Ana'kh. When you look at a page, you encounter about three hundred *Nokta* – meaning spots, or messages. You look at a certain Nokta, and it opens up to thousands of other words and meanings.

The content is huge, but not intimidating, since it opens up in what seems to be multiple screens. Then, you can choose what you are interested in. I was reading along, finding it very easy to understand the pure, traditional Ana'kh, and completely comfortable with the viewing machine, so much so that I no longer noticed wearing it.

I was particularly interested in the creation of humanity, so the book took me to that moment in time. I kept doing this, moving from one Nokta to another, until I decided to move to another subject. I was fascinated by what the book had to offer regarding the dimensions and limitations of the universe. I got the precise information I wanted regarding the question of whether the universe is expanding or shrinking. After that, I wandered into a Nokta regarding the future of humanity. One thing led to another, and I was so totally absorbed, that I did not know if the other three were still with me or not, and certainly did not know how much time passed. Finally, after watching millions of years enfold in front of me, I pulled back with a sigh.

I felt the hand of Dr. Farid on my shoulder, and turned. "Do you know that you have been reading for two days?" he asked, smiling.

"Two days?" I asked, startled. "I did not eat, or drink, or sleep for two full days?"

"Yes," said Dr. Farid. "And you squeezed millions of years into two days. Time to go." I did not feel it. Not a bit of exhaustion or thirst or hunger was caused by this intense study that lasted two days.

On the contrary, I felt as comfortable and refreshed as if I came back from vacation. I mentioned that to Dr. Farid on our way back and he said that this was a common reaction, though some people did feel rather exhausted. Apparently it was an individual reaction. Still, he advised me to go to the hotel and rest."

*** *** ***

NOTES

NOTES

CPSIA information can be obtained
at www.ICGtesting.com
Printed in the USA
BVHW031924190521
607750BV00013B/78